Contents

Construction

Costs/Benefits

PRACTICAL SOLAR HOT WATER

A HOME OWNER'S GUIDE

Kevin McCartney | Brian Ford

VAN NOSTRAND REINHOLD COMPANY
NEW YORK CINCINNATI TORONTO LONDON MELBOURNE

Library of Congress Catalog Card Number 82-13699
ISBN 0-422-26471-2

Printed in Great Britain

Published by Van Nostrand Reinhold Company Inc.
135 West 50th Street, New York, NY 10020

Van Nostrand Reinhold Publishers
1410 Birchmount Road
Scarborough, Ontario M1P 2E7, Canada

16 15 14 13 12 11 10 9 8 7 6 5 4 3 2 1

Library of Congress Cataloging in Publication Data

McCartney, Kevin.
Practical solar hot water.
2nd Edition Originally published: Practical solar heating. Prism Press, 1980.
Includes index.
1. Solar heating — Amateurs' manuals. 2. Solar water heaters—Amateurs' manuals. I. Ford, Brian.
II. Title
RH7413.M39 1982 696'.6 82-13699
ISBN 0-422-26471-2

Appendices

FOREWORD

Domestic use of solar energy is no longer a science fiction proposition. The principles behind solar water heating are as simple as those behind drying clothes in the sun.

That is not to say that constructing and installing a water heating system is quite so simple as hanging clothes on the washing line. Still it is within the scope of the home handyman or woman. And, once installed, it will operate automatically for many years, heating water whenever weather permits.

Even in northern latitudes almost half of the annual domestic hot water requirements can be supplied by the sun without any changes in lifestyle. The exact savings will vary, not only with the weather, but also with the size of the collector, the type of system and the amount of hot water required and the time it is used. Energy savings would be increased if hot water consumption was controlled to coincide with the availability of solar heated water – choosing a sunny day to have a bath for example, and giving the solar tank time to heat up by waiting till mid-day before starting to do the laundry.

Experiments in solar water heating have been carried out for almost a hundred years. Only cost has held up its application in ordinary homes. Now, the greatly increased price for conventional energy sources, reflecting the rapid depletion of fossil fuel resources, has made solar heating economically competitive in many situations.

With the availability of Federal tax credits equivalent to 40% of the cost of a solar system, now is the time to buy solar. Furthermore, there have been considerable advances in plumbing techniques over the past decades. These have made do-it-yourself installation feasible thereby enabling the initial capital cost to be cut by about one third.

For those who would like to do something practical about the energy problem but do not know a thermosyphoning system from a differential temperature controller, this book provides an introduction to the theoretical aspects, descriptions of the hardware involved and a do-it-yourself guide to the construction and installation of solar water heating systems. For those looking for an off-the-shelf solution, who have been confused by the wide variety of products there is a chapter on collectors which will prepare them to meet the salesman of commercially available equipment. Those who are still undecided might be interested in the sections on costs/benefits, subsidized loans and how solar systems are integrated with conventional hot water systems.

1

A SAFE ENERGY SOURCE

The Sun God

Many of our ancestors took a religious view of the sun as the great benefactor and source of all life. This was a soundly based philosophy and formed a solid foundation for a society supported by an agricultural economy in which knowledge of, and reverence for the sun god's nature and movements were closely linked to survival. For then, as now, it was the sun's energy which supported all life — collected by the leaves of green plants, then absorbed as food by the animals.

Mythology however shows that even in ancient times there were men who hoped to enslave the sun for their own ends. Polynesian Indians tell of a hero named Maui who at dawn laid in wait at the Eastern edge of the world, armed with nets and strong ropes to ensnare the rising sun. He is said then to have beaten the sun half to death and as a result the days since have been much cooler.

People of more northerly latitudes are unlikely to thank Maui for the cooler days we have had, and would do well to adopt a more gentle approach to catching the sun. Nowadays piping and sheets of glass have profitably been substituted for ropes and nets. It would be unfortunate though if the Maui-type attitude were retained and only the implements changed. It is after all, the predisposition for indiscriminate exploitation which has brought us to the environmental and cultural crises which face the industrialised world. Energy shortages are only a symptom of this larger problem.

Bearing in mind the warnings ecologists have been giving us throughout the past decade we would perhaps improve our present-day survival chances if we adopted some of the reverence which the followers of Ra, Aten, Phoebus Apollo and other solar deities displayed. Watching the skies, acknowledging our environment, growing increasingly aware of the fantastic swirl of solar energy which sustains the life on our planet, we can begin to solve the problems which confront us.

Atoms and Energy

The attempts of the ancients to explain the wonder of solar energy are no more amazing than the current explanations put foward by scientists.

Every thing we see, feel and touch is made up of tiny particles which we call atoms. Physics teaches us that the atoms themselves are made up of still smaller parts—a nucleus of protons and neutrons surrounded by orbiting electrons. It was discovered that even these can be broken down into still smaller components.

Modern physics paints a weird landscape in which apparently solid matter can be seen to consist largely of voids, populated by miniscule particles which behave at one time like pieces of matter, and then at another time like waves, or bundles of energy.

All matter can be seen to be ultimately a concentration of energy, exquisitely ordered in a limited number of patterns to produce the elements which chemistry uses as building blocks to produce all the familiar compounds we find in our daily lives.

The concentration of energy in the nucleus of an atom is intense. Hence its potential for releasing energy is immense. This has been demonstrated in the atom splitting experiments where atoms are forced to collide with other particles at great speed. This can cause the atom to disintegrate. Most of the particles recombine very quickly to form new atoms but a few sometimes escape from their material form thereby releasing vast quantities of energy.

Nuclear Power Station in the Sky

The same process occurs in the sun. Under conditions of extremely high pressure and temperature, light elements such as hydrogen fuse together to form heavier elements. Such transformations release the torrent of radiation which streams forth in all directions into space.

The flow of energy from the sun is so great that even our tiny planet, safely removed, ninety three million miles from the inferno, intercepts some four thousand million, million kilowatt hours each day. In ten days, we receive more energy from the sun than is contained in all the world's fossil fuel reserves.

The Peaceful Atom

The atomic bombs dropped on Hiroshima and Nagasaki in 1945 demonstrated publicly that mankind had learned to release the energy in the atom. Following the awesome success of the war technologists, in demonstrating their ability to kill and destroy at a rate previously undreamed of, the technically advanced nations began to devote the bulk of their energy research funds into nuclear power.

In the 1950's, promises of cheap, unlimited supplies of nuclear power within two decades made nuclear energy the cornerstone of all the developed world's future energy policies and the "peaceful atom" was heralded as the basis for a new era of prosperity.

Thirty years later, the energy crises of the seventies and the spiralling rise in costs have served to underline the failure of the earlier promises. Technical difficulties have meant that the fast breeder reactor has still not even been demonstrated on a commercial scale. The fast breeder through its ability to use uranium fuel more efficiently, is the only type of nuclear reactor which might be considered to have any long term future.

The Nuclear Threat

Meanwhile the thermal fission reactors which have been commissioned continue to produce radioactive waste with no satisfactory solution yet devised for safeguarding these materials some of which will remain a health hazard for twenty thousand years. The Three Mile Island Incident which shook Harrisburg in Pennsylvania, has revived fears of a disastrous nuclear reactor accident of the type which the industry's statisticians have been trying to persuade us will never happen. Even without major accidents occurring, so-called acceptable leakages from nuclear power plants will raise the background radiation level to which the whole population is subjected. American doctors Tamplin and Gofman have estimated that this could lead to an increase of 32 thousand in the annual U.S. cancer death toll.

Apart from the threat to health, proliferation of atomic fuels such as plutonium will form a threat to peace as they constitute the raw material for the fabrication of atomic weapons. The danger of terrorist abduction during the transportation of such materials will increase the necessity for security operations which are likely to further infringe upon individual liberties.

The Solar Alternative

The establishment of an energetic basis for our society which is sustainable long into the future and does not constitute a threat to the very population it is designed to support, must be a primary concern. The first step is to reduce our energy requirements. Given the present level of inefficiency of fuel usage in our homes, workplaces and transport systems, it is fairly simple, and also financially profitable, to dramatically cut our energy consumption without lowering our living standards. But we must also begin now to provide for the future. Due to the long lead-in period required to enable a new source of energy to contribute significantly on national energy balance sheets, it is essential that we start immediately to deploy the energy technologies which we intend to have supplying our needs at the turn of the century.

Solar energy offers the only opportunity for a safe and reliable energy future, and we do not need to wait for the utility companies and the international corporations to decide that for us. By choosing the solar alternative for providing heat in our homes, offices, factories and community facilities, we can positively contribute to the creation of a peaceful and secure future.

2

UTILISATION OF SOLAR ENERGY

Deliberate indirect utilisation of solar energy began with agriculture. Our farming forefathers selected, protected and nourished certain plant species which they considered useful. These plants converted a small proportion of the solar energy falling on their leaves into chemical energy which held together the compounds which served as food or clothing for humans. This conversion of energy in the plant is called photosynthesis, and is carried out by chlorophyll—the substance which gives plants their greenness. Sunlight enables chlorophyll to fix carbon dioxide from the air, and water from the soil to produce carbohydrates. These carbohydrates constitute a biological energy source for the plant, or for an animal which might eat the plant. All life depends on photosynthesis. All life energy has been transmitted from the sun via plants.

Five Star Solar Energy

When our ancestors learned to handle fire, there began a new line of development in the indirect utilisation of solar energy. This has culminated in rockets carrying men to the moon, centrally heated houses and all manner of 'horseless carriages'. Since setting light to a few twigs and releasing the solar energy which a tree had collected and stored over a period of years, there has been a fascination with the potential for discharging large quantities of concentrated energy.

The discovery of the fossil fuels, coal and petroleum, has fundamentally influenced historical development.

With a flick of a switch, modern man turns on a fire and spills out solar energy which shone down on the lush vegetation of the Carboniferous forests several hundred million years ago. It became buried with the dead plants which were eventually transformed under great pressure from vegetable waste into the mineral deposits which we mine today.

Our present day society has become addicted to this energy concentrate—a candybar diet of instant energy. Exploitation of the energy deposits has been so rapid however that after only a couple of centuries of feasting, the stockpile which took millions of years to accumulate, will soon be depleted. Long before the fuel reserves are actually exhausted of course, price increases will make them inaccess-

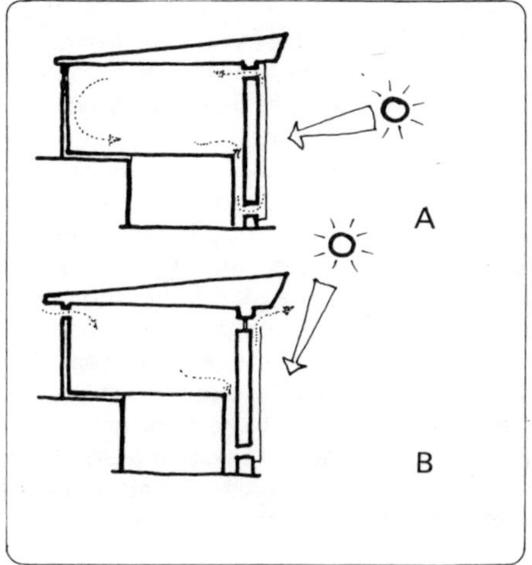

2.5 Trombe Solar Wall—black painted south wall covered with glass.
A Winter heating mode.
B Summer cooling mode.

2.6 Simple Solar Water Heating System—Flat plate collectors with hot water storage tank.

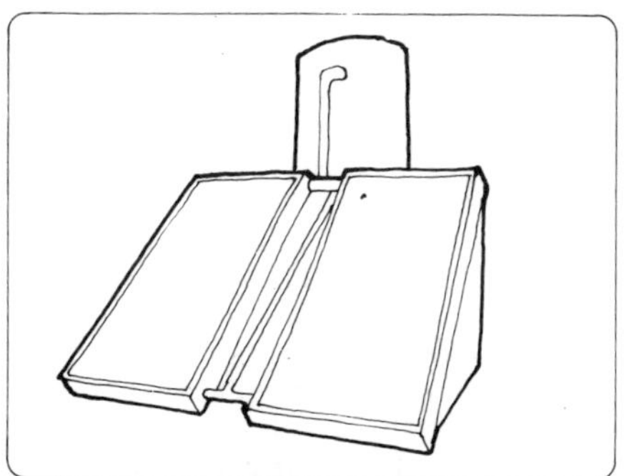

edge of the troughs which lead it to a storage tank.

Solar Cooling

To use the sun for cooling buildings must be one of the most resourceful applications of solar energy. The simplest method used is to create a strong convection current in the air which will draw a draught of fresh air through the building. This has been done by covering the south facade of the building with glass and leaving an opening at the top. Air between the glass and the wall will rise when heated and will escape through the opening. If ducts are made through the base of the wall, the rising current of air will suck air from the interior through them and hence draw fresh air into the building from windows in the shaded north side. Where clear night sky conditions prevail, nocturnal cooling can be achieved through radiation from a flat roof. Cooling effectiveness can be increased further by spraying or flooding the roof with water and enabling heat to escape through evaporation.

More complex solar air conditioning systems have been installed in experimental houses. These are not yet economically competitive with conventional refrigeration installations. They require high temperature solar collectors to power a heat engine which then drives a compressor in a standard cooler, or they are based on the absorption refrigeration cycle as in a gas fridge.

Flat Plate Solar Collectors

The solar installations which will most rapidly pay back the cost of their construction, in the form of reduction in fuel bills, are those which serve a requirement for low grade heat, i.e. where temperatures below 180°F (80°C) are sufficient. Such temperatures can be attained in simple flat plate solar absorbers, insulated in weatherproof casings, glazed and mounted to face the sun's midday position. They have no need of complex tracking mechanisms and they can also make use of some of the scattered radiation which is so abundant on bright but cloudy days. Flat plate solar collectors can be used to heat water for washing, swimming pools, some industrial processes and for room heating.

The basic flat plate solar system has three functions: collection, circulation and storage. There follows a chapter on each. Briefly one can say that the collection takes place by absorbing radiation from the sun with a black coating. Once absorbed the radiant energy converts to heat. The circulation of a fluid through the absorbing plate allows the heat to be withdrawn. If the fluid is passed into a tank, this then serves as a storage medium which can store the heat until such time as it is required.

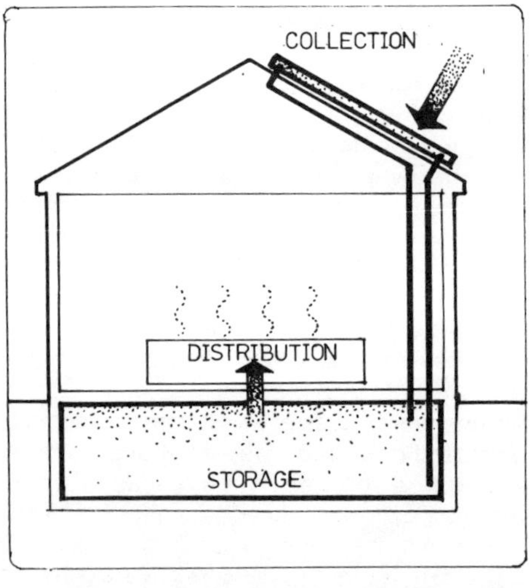

2.7 Solar Space Heating System—Flat plate collectors mounted on roof with large underground storage tank.

Space Heating

The greatest difficulty in using solar energy for space heating arises from the obvious fact that the need for heating is greatest when the sun is absent. A solar space heating system therefore has to have an auxiliary heating system to maintain comfortable room temperatures during long periods without sun, or it must have a huge storage tank which will enable heat collected during summer to be held over for use during winter. Such a storage tank might be as large as the house it served and is therefore a very expensive item. Most systems which have been built make a compromise and include a storage capacity of 1–2000 gallons (4—8 cubic metres). When topped up with hot water, this is large enough to supply a house's heating requirements for several days. Some form of auxiliary heating would be required with such systems except in very mild climates.

A more economic means of using solar energy to reduce a building's heating requirements is to exploit the greenhouse effect. Gardeners are familiar with the fact that a glass enclosure can become very warm whenever there is sun, even on cold winter days. Greenhouses can be built on the south side of houses providing an area for food production as well as a source of warm ventilation air.

Alternatively, the south wall can be glazed

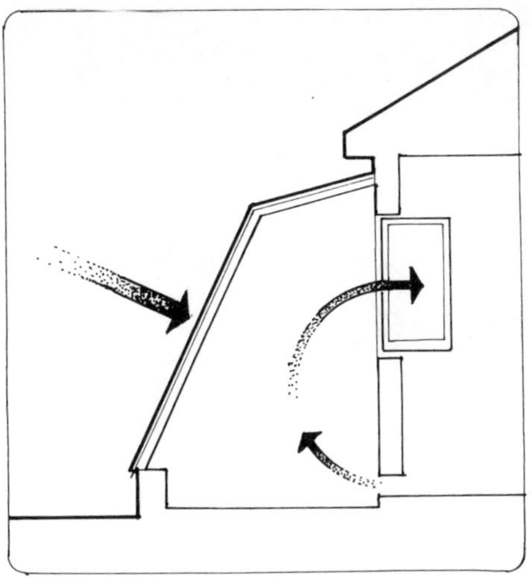

2.8 Passive Solar Heating—a lean-to greenhouse on the south wall supplies warm air to the building.

with only a small space between glass and wall, as in the Trombe Wall described in the section on cooling. This same construction can serve to heat the building by closing the opening at the top of the glass and allowing the rising warm air to enter the building through ducts in the top of the wall. If the duct is closed, there will be no air circulation to carry the heat away and heat will be absorbed, therefore, by the wall itself. If it is a thick solid wall, say a foot (300mm) of concrete, the heat will take several hours to pass through to the internal surface. Hence the wall would not begin to radiate heat into the room until evening at which time it is most likely to be needed.

The simplest solar collector of all is a south facing window. Depending upon their size, they can contribute some 10–20% to a house's heating requirements. Heavy curtains or insulating shutters should be used in conjunction with any large windows in order to reduce night time heat loss.

Solar Water Heating

A solar water heating system will for many be the first step in going solar. Hot water is required all year round. Water heating systems therefore can usefully operate during the summer period when the supply of energy is at a maximum. This in turn means that the installed equipment can pay for itself, in the form of reduced fuel bills, much more quickly. Solar water heating systems are also smaller and simpler than mechanical solar space heating systems. Thus they give the householder an opportunity to dip a testing toe in the water before diving in at the deep end with a big investment in a whole house heating installation.

This chapter has outlined just a few of the practical techniques for utilising the sun's energy. The chapters that follow will concentrate on solar water heating. They will provide enough information to enable you to go out and make sensible choices in buying and installing a system in your home. The do-it-yourself enthusiast will also find the construction details and practical tips required to facilitate the building of a complete system.

3

BASIC PRINCIPLES

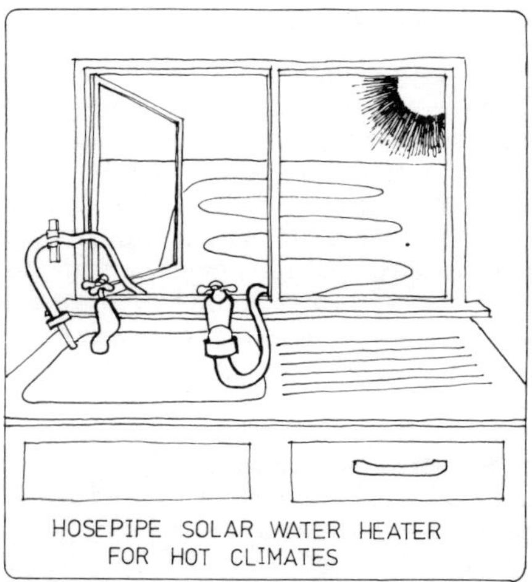

HOSEPIPE SOLAR WATER HEATER FOR HOT CLIMATES

Collecting solar energy can be very simple. In the sunniest parts of the world some householders simply snake a black hose pipe in a long loop from their kitchen window, round the yard and back to the sink. And that is all that is necessary to heat their water in the summer.

Where the climate is less favourable, solar collection systems need to be somewhat more sophisticated. They have to put to use as much of the available sun as is practicably possible, collecting it more efficiently and having the capacity to store it for at least the few hours between a sunny morning and the period of peak demand in the evening.

To understand how such improvements can be brought about, it is helpful to consider how energy, particularly heat, moves about and how radiation from the sun interacts with objects which lie in its path.

Heat Transfer

Heat can be transferred between objects in three different ways: conduction, convection and radiation. Conduction takes place when heat travels from one molecule to another through direct physical contact. It is through conduction that the handle of a pan can become hot when it is sitting on the stove. Convection takes place only in fluids. Liquids and gases expand when heated. This means that each unit volume of the material becomes relatively lighter. The heated fluid therefore tends to rise above surrounding cooler fluids. In this way heat is lifted away from a hot object in what we call convection currents. This can be seen in the clouds of steam which rise above a heated pan of water.

Radiant Energy

Radiant energy, unlike conducted and convected energy, does not require a physical medium through which to flow. It can pass through a vacuum carrying energy from a hot object to a cooler one. Hence solar energy can cross the gulf of space to bring warmth to the earth. It moves in waves like the ripples which spread out from a pebble dropped in a pond. Very hot objects radiate energy in such a way that there is a very short distance between the crest of each wave, i.e. in short wavelengths. Cooler objects radiate in longer wavelengths. As will be explained later, this difference in the wavelength of radiated energy gives rise to the 'greenhouse effect' which can be of great assistance to systems designed to collect solar energy.

Colliding with a Sunbeam

When an object lies in the path of radiant energy three things can happen: energy can be transmitted through the object, reflected away, or absorbed by it. The reaction depends upon the physical characteristics of the object, the wavelength of the radiation and the angle of collision.

Transmission

Transparent materials like glass are good transmitters of solar energy. If solar radiation approaches a sheet of glass at right angles, nearly all the energy will pass through it. Most of the energy which is not transmitted is reflected away. The more oblique the angle between the glass and the sun's rays, the greater the proportion of energy which will be reflected. This explains the dazzling light which can be seen reflected from south-facing windows when the sun is sinking in the west. This characteristic also indicates that the amount of energy entering a glass covered solar collector can be maximised by tilting it so as to be perpendicular to the sun's rays.

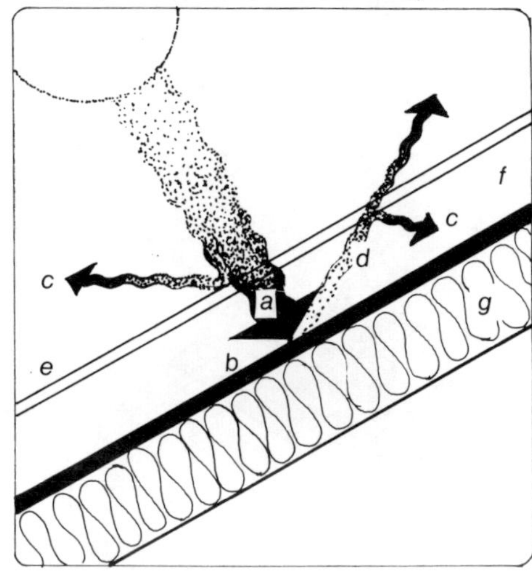

3.2 a) Transmission b) absorption c) reflection & d) re-radiation in a solar collector e) glass f) absorber g)insulation

Wavelength and Color

Solar radiation consists of many different wavelengths. Most of the sun's energy arrives in very short wavelengths which scientists measure in units called microns. There are almost 25,000 microns in one inch. Surprisingly enough our eyes are able to distinguish between wavelengths whose lengths are different only by a fraction of a micron. Few of us though, describe what we see in terms of microns. Rather we use words like violet, indigo and blue for the shortest wavelengths and red, orange and yellow for the longer wavelengths. Not all solar radiation is visible however. The infra-red component has a wavelength too long for our eyes to register, and the ultraviolet is too short.

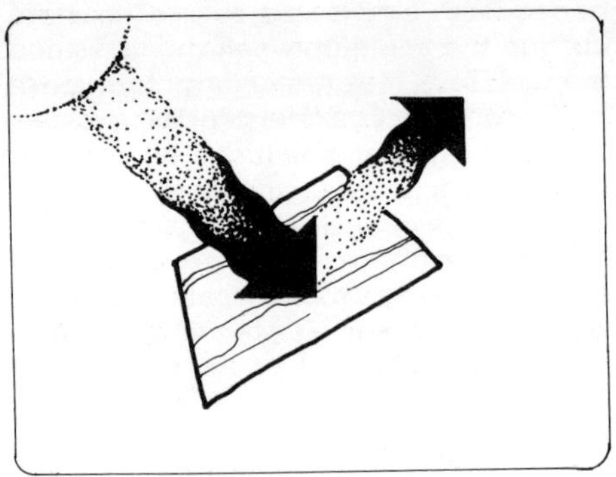

3.3 *Shiny surfaces reflect radiation*

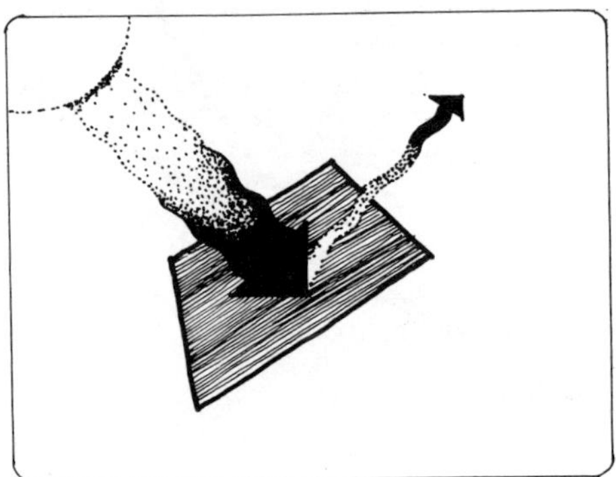

3.4 *Dark surfaces absorb radiation*

Reflection and Absorption

All the colours we see in a landscape are components of solar radiation being reflected in accordance with their wavelength by different materials. The difference between a yellow daffodil and a red rose in terms of their interaction with sunlight is that the daffodil absorbs most of the radiation with the exception of those wavelengths which we perceive as yellow. These are reflected from the petals and into our eyes The rose on the other hand absorbs the yellow wavelengths but reflects the red. Another rose, with a different chemical structure might reflect nearly all the different wavelengths. When all the colors are reflected together, we see white. In the opposite case, where nearly all the wavelengths are absorbed, we see black objects.

Knowing this, we can actually increase the amount of energy an object absorbs simply by giving it a black surface coating.

When radiant energy is absorbed it changes into heat. One might imagine the solar rays, having beamed ninety-three million miles through space, suddenly crashing into an absorbing object. The 'shock' waves set the electrons in the molecules of the absorber vibrating more rapidly. If this shaking up of the electrons is violent enough, we can actually sense their movement as warmth, and measure the change as a rise in temperature.

Hot and Cold Cars

The effect of different colored surface finishes on energy absorption can be felt. If you see a white automobile parked beside a darker one in the sun, lay a hand on the bodywork and the temperature difference should become obvious.

Re-Radiation

Once the temperature of the absorbing object begins to rise, it also begins to radiate heat. If the absorbing object is intended to function as a collector of solar energy, this re-radiation must be kept to a minimum. This can be done in two ways:

a) exploiting the greenhouse effect.

b) maintaining a low operating temperature.

The Greenhouse Effect

The greenhouse is like a heat trap. It works because glass, and some plastic materials, act like radiation filters. They transmit a large proportion of solar energy which is mostly short wave radiation, but they do not transmit the longer waves which are reradiated from objects heated in the sun. Thus if an absorbing object is covered by a sheet of glass, heat will build up and we will obtain the higher temperatures which gardeners experience in their greenhouses.

Operating Temperature

If the temperature of the absorbing object is kept low by rapidly extracting heat and transferring it to a store, the amount of energy re-radiated will be correspondingly lower.

When people hear that you are working with solar energy, one of the first questions they will ask you is, "What kind of temperatures do you get?"

I have measured temperatures over boiling point in single glazed flat plate solar collectors when the water is not circulating, and I know of others who have measured temperatures as high as 365°F (185°C) in solar collectors which have been emptied of their water content. When a collector is connected to a storage tank and water circulated through it, its temperature is more likely to be somewhere below 180°F (80°C).

All this talk of boiling water in solar collectors however can be something of a red herring. High temperatures do not indicate an efficient system. Ideally of course one would like to end the day with a tank full of water heated to the temperature at which it is used (115–135°F (46–57°C) in domestic situations). It will be easier to achieve this however, if we operate the collector with only a small temperature difference between the inlet and the outlet.

A solar collector operating at 80°F (27°C) is collecting energy more efficiently than a similar collector operating at 140°F (60°C). At first this seems ridiculous, but after a little thought it makes sense.

The major source of inefficiency in a solar collector is the heat losses which occur after energy is absorbed. The hotter an object becomes, the greater will be the rate at which it loses heat. A collector operating at the lowest useful temperature will therefore function at its highest efficiency.

The operating temperature can be changed by varying the rate at which water flows through the collectors and by altering the connecting patterns in an array of collectors.

The flow rate can be increased, and hence the operating temperature lowered by reducing the circulation resistance, through using larger diameter piping and minimising the length of the pipe runs and the number of bends. With a variable speed pump, the flow rate can be altered at the turn of a switch.

The effect of different connection patterns between solar collectors is discussed in detail in chapter 10.

Tracking the Sun

The maximum amount of radiation is intercepted if the absorber is at right angles to the rays. The more oblique the angle, the more energy will be missed—that is to say the radiation actually falling on the object will be less intense.

In an ideal solar collector, we might want the absorber to track the sun on its changing path across the sky, its tilt varying as the sun rises to, and falls from its midsummer zenith. The mechanism for such a tracking process would be unjustifiably expensive and complex. When using flat plate absorbers which do not rely upon any focussing of the sun's rays, satisfactory results are obtained with a stationary mounting if it provides a tilt and orientation within the limits suggested below.

Tilt Angle

The further we go from the equator, the steeper we have to tilt a solar collector in order to keep it at right angles to the rays of the midday sun. We call the angle between the collector surface and the horizontal the *tilt angle*. The rule of thumb is to set the tilt angle equal to the latitude of the site where you are installing the system. Deviations of plus or minus 10 degrees will not cause any significant reduction in the amount of sunshine you capture.

In cloudy climates, the ideal tilt might be somewhat flatter. In London, England for example, at a latitude of 52°N, the collector which would receive the most energy throughout the year would have a tilt of

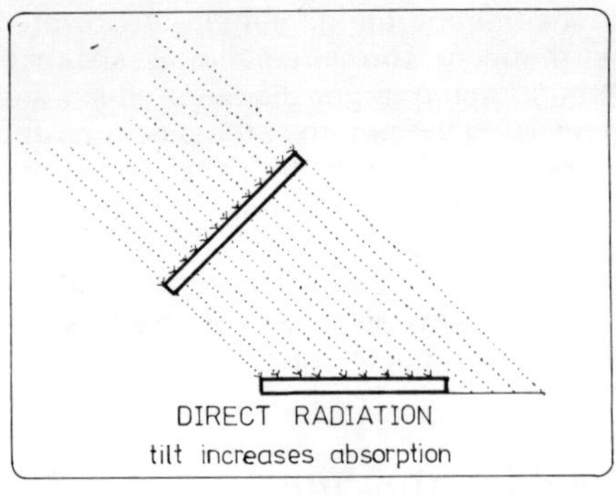

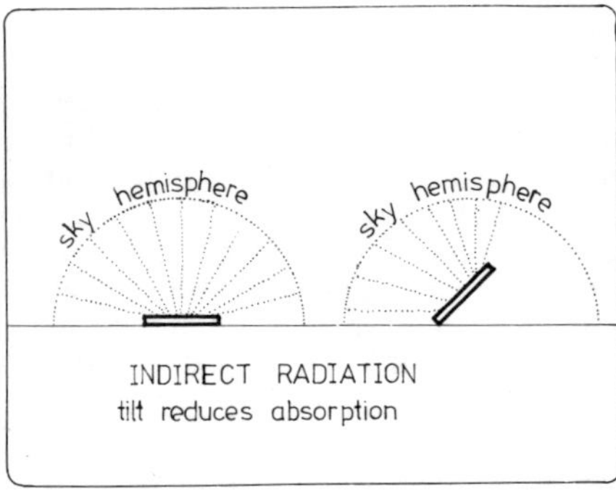

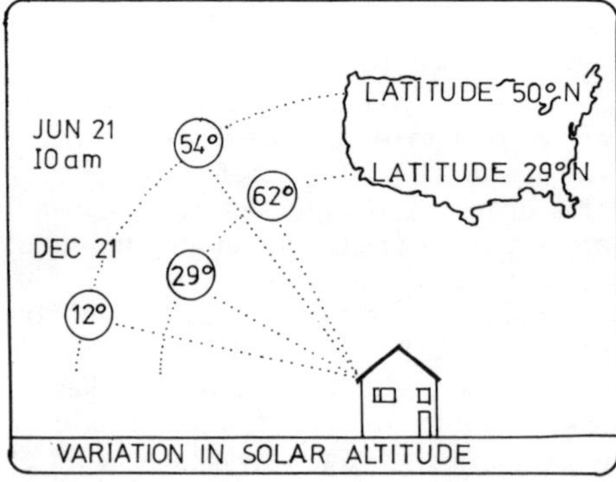

only 34°. This discrepancy is due to the fact that almost half the solar energy received comes not directly from the sun, but is diffused by the cloud cover. This diffuse radiation emanates from the whole sky hemisphere. The ideal angle therefore for a collector aimed at diffuse radiation would be horizontal. This accounts for London's annual optimum tilt of 34° being much flatter than its latitude would lead one to expect. When we are considering an application like space heating the demand of the system will be at a maximum in winter when the sun is low in the sky. In such cases a tilt angle greater than the latitude angle would be appropriate. Similarly, a steep angle would be appropriate in a 100% self-sufficient house where normal heating is not available.

In such a case it is necessary to optimize the system for the winter months when there is much less sun available. In ordinary housing however, where the intention is to maximize the saving of gas, oil or electricity for water heating, it is best to optimise the system's performance for autumn and spring.

This discussion of optimum angles may be irrelevant if you are hoping to install solar collectors on an existing pitched roof. In this case, altering the pitch or constructing special supports is unlikely to be worthwhile. In some regions, the ability to shed a snow load will be a critical factor in the choice of tilt angle. In such cases, following local conventional roof slope angles will be a practical guide. Reductions in energy received due to non-optimal tilt angles, can be compensated for by increasing the collection area.

Orientation Angle

Although it is obviously desirable to have the solar absorber facing south, it can in fact be orientated as much as 20 degrees east or west of south with only a negligible reduction in the amount of energy received. Beyond 30 degrees away from south the reduction becomes more marked, particularly in the winter months.

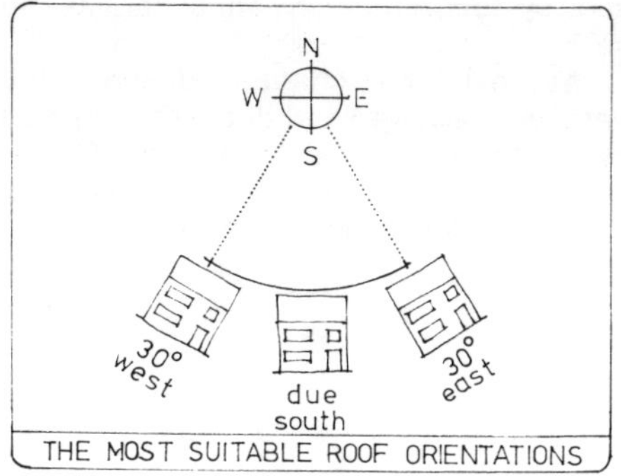

THE MOST SUITABLE ROOF ORIENTATIONS

For absorbers with steep angles of tilt, the losses encountered when facing away from south will be slightly greater. Facing away from the sun's midday location, they rely much more on diffuse radiation. Being so steeply tilted however they turn their back on as much as half the radiating sky hemisphere. A southerly orientation is therefore more critical with vertically mounted absorbers than those mounted on pitched roofs.

Surface Area

Solar energy can be considered as an infinitely large waterfall pouring itself across the earth's surface. If we are attempting to intercept this flow, it is obvious that the larger the area of the interceptor, the greater will be the catchment. So it is with solar absorbers.

There is though a limit to the amount of surface area which will be useful in any particular situation. This is because a large area which would prove useful in winter, would be largely redundant in summer, producing more heat than could be used, perhaps even requiring cooling. To build too small a collector area can also carry an economic penalty due to the high cost of ancillary equipment such as tanks and pipes and pumps. Typical family hot water systems have 30–80 sq. ft. (3–8 sq.m.) of collectors. Chapter 13 will help you determine a suitable size of collector taking account of climatic conditions and hot water requirements, and will also help you analyse the financial implications of your choice.

Surface Area & Storage

Whilst a large surface area is beneficial for the collection of energy, it is a positive disadvantage for the storage of energy. Heat is lost in proportion to the surface area of the storage vessel. A flat plate of large surface area which is an ideal form for an absorber is the worst form for storage.

A similar contrariness between the functions of collection and storage arises with respect to the characteristic of thermal mass. To optimize both functions it is necessary to separate them.

Thermal Mass & Reaction Time

Thermal mass is a term used to describe an object's capacity for storing heat. Matter with a high thermal mass may therefore be useful for storing large quantities of heat in a small volume. It will however require a comparatively long period of exposure to sunshine in order to experience any considerable rise in temperature. Matter with a high thermal mass is typically dense, like water and granite. If as suggested the functions of collection and storage are separated low thermal mass in the absorber can be advantageous, as this would allow a more rapid temperature rise. In a solar collector system, it is the temperature rise in the absorber which begins the circulation process which transports absorbed energy into the storage vessel. A low thermal mass will therefore mean a quick reaction time.

In sunny conditions this property is not so important. Once circulation has begun, it will continue throughout the day. In conditions where periods of intense radiation are followed by dull spells, a collector of high thermal mass might have been absorbing energy for several hours and be just reaching the temperature where circulation would begin when a layer of heavy cloud causes a sudden drop in the level of radiation. If the surrounding air temperature is low the temperature would gradually fall again as heat is lost to the environment.

Conclusion

This chapter has introduced some of the factors which underlie the operation of a solar system, but which cannot normally be seen.

The next three chapters describe the hardware required and detail the function of the three major components of solar water heating installation: the collectors, the circulation system and the heat store.

4

SOLAR COLLECTORS

Solar Collectors are to a solar water heating system what a boiler is to a conventional system. It is in the collectors that the water is heated. The water is circulated through the collector so that solar heat, trapped by an absorber plate, might be transferred to it by physical contact.

The collector consists of four elements:

1) absorption plate
2) translucent cover
3) insulation
4) casing

The case is in the form of an open-top, shallow box, insulated on the bottom and sides. Within lies the absorber, and above it, like a window, is the translucent cover. If the collector is integrated in the roof of a building, the roof rafters may take the place of the casing. In applications where the water does not need to be heated more than 10°F (6°C) above the air temperature and the collector site is not exposed to strong winds, an absorber plate alone, uninsulated and uncovered might prove adequate.

How It Works

When solar radiation reaches the cover, three things happen. A small quantity of the energy is absorbed by the cover itself; some is reflected away; the remainder is transmitted through to the absorber plate below. Most of this transmitted energy will be absorbed and transformed into heat which is conducted through the plate to the water channels which are either integral or bonded to it. If water is then circulated through the absorber plate, the collected energy can be extracted.

Unfortunately not all of the absorbed energy can be extracted. For as soon as the absorber plate's temperature begins to rise, it will also begin to lose heat to its surroundings. Heat will be conducted through the casing with which it is in contact. The air in contact with the plate will become heated and rise in a convective current, thus carrying away more heat. Finally the plate will radiate energy outward at a rate which will increase with its temperature.

Efficiency

When people talk about the efficiency of a collector they are referring to the quantity of energy extracted by the circulating stream of water expressed as a fraction of the total amount of solar energy falling on the cover of the collector. Hence, on a day when 14,000 Btu (4 kWh.) of solar energy was received on the cover, and the collector contributed 7,000 Btu (2 kWh.) towards heating water, the collector can be said to have operated at 50% efficiency. The efficiency of a collector is not a constant value. It changes with air temperature, wind velocity, solar intensity and the water temperature in the collector. Domestic solar water heaters usually operate at efficiencies between 30% and 50%.

From the description of how the collector works, it can be seen that several things might be done to ensure that the efficiency is as high as possible. The energy gains must be maximized and the heat losses minimized.

In other words we should maximize:

a) transmission through the cover
b) absorption by the plate surface
c) heat transfer from the absorber to the water

At the same time we should minimize:

d) heat conduction through the casing
e) Convection currents on the surface of the absorber
f) re-radiation losses from the plate

Transmission through the cover

The proportion of solar energy which passes through the cover depends upon the physical properties of the material used, the cleanliness of its surface and the angle at which the rays strike.

Glass

A clean sheet of ordinary window glass transmits about 84% of the energy it receives. Sometimes a special glass with a low iron content is selected because this will absorb less radiation. The actual increase in transmitted energy however is not large enough to justify the increased cost of the glass. It may be worth noting however that a high iron content, and hence a lower transmissivity, is characterised by a greenish tint in the glass.

The biggest problem with glass is that it breaks easily. Double strength glass might therefore be an appropriate choice. If the hazards of breakage are very great, or where heavy hailstorms occur, tempered glass is desirable.

Plastic Covers

Many plastics are capable of transmitting as much, or even more, energy than glass. Most are prone to deterioration in their transmissivity however, when exposed to the ultraviolet light of the sun. The low temperatures at which many plastics soften and warm is also a problem when considering their use in solar systems. One more disadvantage is that some plastic glazing materials transmit not only the sunlight, but also the long wave energy reradiated from the solar absorber when it becomes heated. This weakens the heat trapping greenhouse effect described in the last chapter.

Plastics do have some advantages though. They are lighter which can be a real boon when carrying solar collectors onto the roof. Even more important, they will not shatter the way untempered glass sometimes does.

Thin film plastics can also be cheaper than glass. The most suitable film glazing for solar collectors is 0.004 inch thick Tedlar, a polyvinyl fluoride (PVF) which should last 5—7 years but may become brittle sooner if subjected to prolonged high temperatures. Care is needed in installing plastic films. They should be stretched tight to prevent them flapping in the wind. Apart from making an annoying noise, this would increase convective heat losses. Special attention is also needed to protect them from tearing due to rough handling. Another useful plastic film is Teflon, a fluorocarbon. This can withstand very high temperatures and transmits up to 96% of the sunlight falling on it. It comes in a thickness of only 0.001 inch which means that it cannot really withstand outdoor weather conditions and is suitable only as an inner layer in a double glazed collector.

The thicker sheets of plastic are easier to install and more durable. The most common type is fibreglass reinforced polyester (FRP). The most widely available grades however cloud up rather quickly under strong sunshine. Ultraviolet resistant grades are available such as Kalwall's Sun-Lite and Filon's Tedlar clad FRP. Flat sheets of FRP are very flexible and therefore need more support than glass. Corrugated sheets are more rigid but have greater heat losses due to their increased surface area. The transmissivity of FRP is permanently lowered by exposure to high temperatures and FRP covers are not recommended for trickle or open type collectors (see below).

Double Glazing

Two layers of glass are sometimes used in order to reduce the heat losses from the absorber. This however will also reduce the amount of solar energy entering the collector. A double layer of ⅛" (4mm) glass will transmit only 71% of the radiation it receives.

Double glazing is only beneficial therefore in circumstances where heat losses are particularly high—in very cold climates for example.

If insulating glass units are used, they must be made from tempered glass, otherwise the thermal stresses experienced by the inner layer are likely to cause breakage.

Dirt

Dust and grease gathering on the cover will reduce the amount of energy transmitted. This problem will be minimised by the use of glass due to its hard smooth surface. Furthermore a steep mounting angle will assist in the cleansing effect of rainfall. Even with a tilted sheet of glass however the transmissivity may drop by about 20% if it is not cleaned regularly. The exact amount will depend on the amount of dust and pollutants in the local atmosphere.

Absorption by the Plate Surface

The absorption of radiation by the plate is increased by coating it with a matt black surface. So treated it will absorb 80-98% of the radiation reaching it. The surface finish should be as thin as possible so as not to form a barrier to the heat flowing from the exterior to the interior where it will be transferred to the circulating stream of water. It is important that the coating is not subject to outgassing at high temperatures as this would give rise to a tarry deposit on the underside of the glazing.

Heat Transfer from Absorber to Water

Heat transfer at this point can be assisted by constructing the absorber plate from materials which are good thermal conductors and by maximising the contact surface area between the water and the water channel, and between the water channel and the absorber plate.

It is principally in the provision for conducting heat from the absorber into the circulating water that absorber plate designs differ. There are three main types:

1) Trickle or open trough
2) Tube-and-sheet
3) Sandwich

Trickle Absorber

This is the simplest and cheapest type of solar absorber. It consists of a sheet of corrugated roof decking with a perforated water feed pipe running along its upper edge, and a gutter pipe along its lower edge. The holes in the upper pipe, the sparge pipe, are situated so that water pours into the gulleys formed by the concave corrugations and then trickles down them to be collected again by the collector pipe at the bottom. In this way water can remove heat which the metal panel has absorbed during exposure to the sun.

Such an absorber should be able to efficiently conduct heat from the convex corrugations or ridges, to the concave corrugations where the water flows. To achieve this, the corrugations should be closely spaced and be good thermal conductors. Of the commonly available corrugated mater-

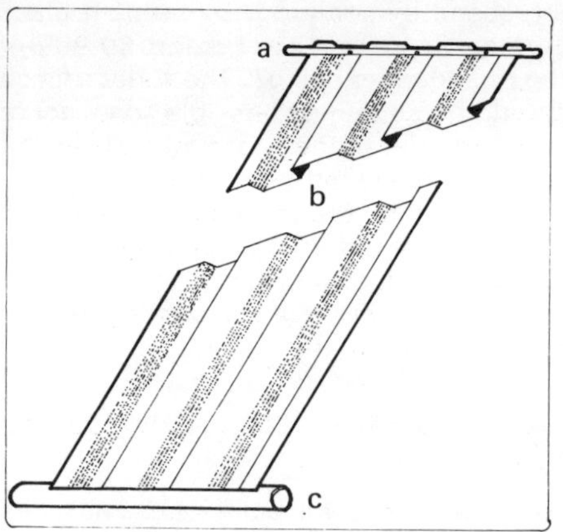

4.1 Trickle Absorber: a) perforated feed pipe; b) corrugated sheet; c) gutter pipe.

ials, aluminium is the best choice. Its thermal conductivity is four times greater than that of mild steel. This means that a steel sheet would have to have its corrugations spaced at a quarter the distance of that on an aluminium panel, or it would have to be four times as thick, in order to match the performance of the aluminium in heat transfer.

The surface contact area can be increased by adding detergent to the water. This will decrease the water surface tension which tends to hold the water stream together, preventing it from spreading laterally across the corrugated channels.

A long lasting and robust coating is essential for this type of absorber due to the harsh environment which it must withstand: the corrosive effect of air and water combined and dry temperatures which may rise to almost 400°F (200°C) causing expansion and rapid contraction when water begins to circulate. A baked-on finish is probably the safest choice.

Having said that the trickle absorber is the cheapest, it is worth pointing out that it does have several disadvantages when compared to those with sealed waterways. It is less efficient because heat is lost when water running down the open channels evaporates. This evaporation causes a second problem in that it condenses on the comparatively cold surface of the glass cover, misting it and thereby reducing the amount of solar radiation which can enter the collector. Precautions must be taken to ensure that the condensation is not able to cause any damage to the roof on which the absorber is mounted. Dust and dirt is likely to enter the system so that even with filtration of the circulating water, access for cleaning the perforated pipe must be possible in case of blockages. Finally, there is more difficulty in accurately controlling the pump for this type of absorber due to the large fluctuation between the plate's wet and dry temperatures.

Tube and Sheet Absorbers

With tube-and-sheet absorbers, care must be exercised in the choice of materials, the spacing of the tubes and the bond between the tube and sheet. The tube configuration may also have bearing upon the type of system in which the absorber can be used. A serpentine zigzag configuration, for example, may offer excessive resistance to flow in a gravity circulation system. On the other hand a grid of parallel pipes between two header pipes might give problems due to air blockages if used in a draindown system.

4.2 Serpentine or Zig-Zag Tube-and-Sheet Absorber.

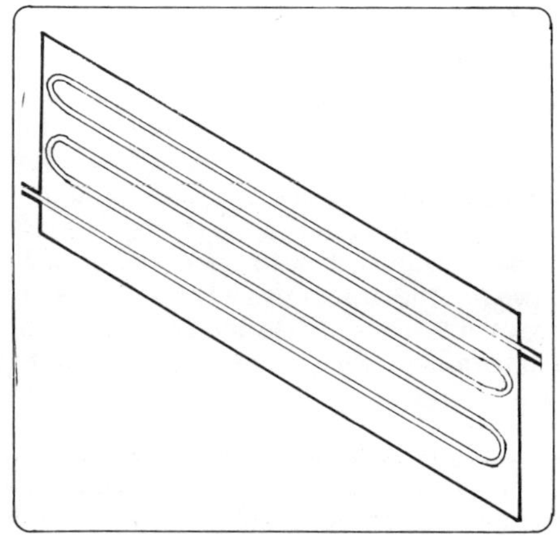

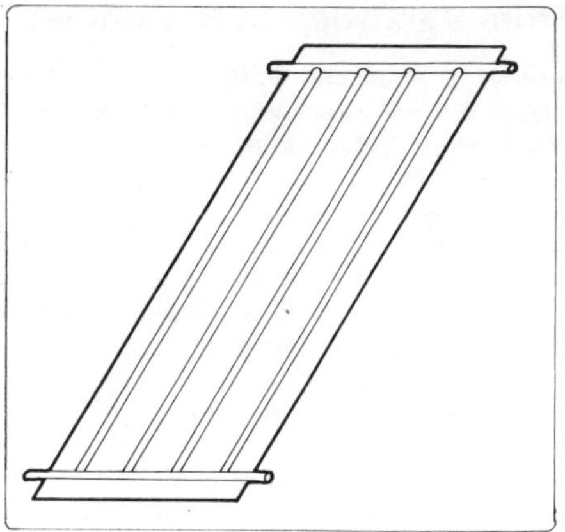

4.3 Tube-and-sheet Absorber with Parallel Grid Configuration.

Tube-and-sheet absorbers, like the trickle type, have to transfer heat laterally, conducting energy collected on the absorbing sheet across to the water filled tubes. For this reason only good thermal conductors can be used. Unfortunately there seems to be some connection between high thermal conductivity and cost. The best heat conductors are materials like gold and silver. Of the commonly available materials copper is the best. It is almost twice as good as aluminium and nearly eight times better than mild steel. Copper is also the most expensive of the three. The ability of aluminium and steel to conduct the absorbed heat to the tubes can be improved by bringing the tubes closer together and by using thicker sheets. The thicker the sheet, the less the resistance to heat flow across it. Typical spacings and thicknesses for the different metals are as shown:

material	**thickness**		**spacing**	
	inches	**(mm.)**	**inches**	**(mm.)**
copper	0.01	(0.25)	5½	(138)
aluminium	0.02	(0.5)	5½	(138)
steel	0.04	(1.0)	4	(100)

When seeking to make a good thermal bond between the tube and the sheet, the first step is to maximise the contact surface area. This can be done by forming the sheet round the tube or by deforming the tube to create a flatter contact surface. If the absorber is all copper, the tube and sheet should be soldered together. With other materials, some form of mechanical bond, clamping or clipping, perhaps in conjunction with a thermal paste, is necessary. Clamping systems should be examined to ensure that they do provide a tight contact along the whole length of the tube.

Sandwich Absorbers

In the sandwich type, water is 'spread' between two sheets of material, the upper of which forms the solar absorbing surface.

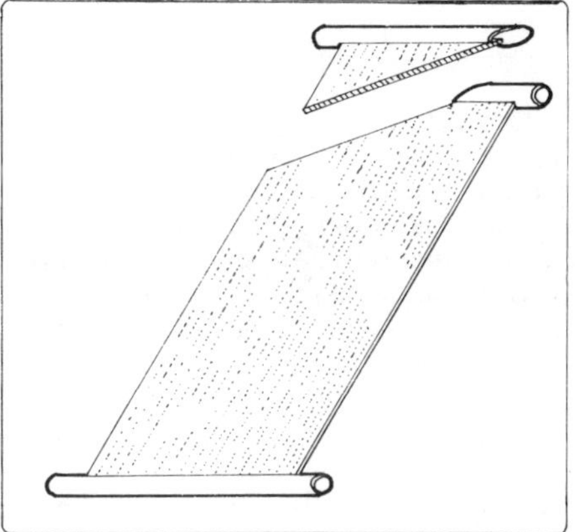

4.4 Plastic Wafer Sandwich

In this case, water is in contact with the whole, or nearly the whole, of the absorbing surface. Hence heat only has to travel through the thickness of the absorber plate wall to reach the water. With this type then, the thermal conductivity of the materials used is not so important as long as they are not very thick. Hence one finds solar absorbers of the sandwich type being manufactured from plastics such as polypropylene. A simply constructed sandwich type of absorber can be made by forming a flat coil of ⅜" (10 mm) copper tube. This is a rather expensive use of copper however.

When more conductive materials are used they are sometimes bonded together in such a way that water is no longer in direct

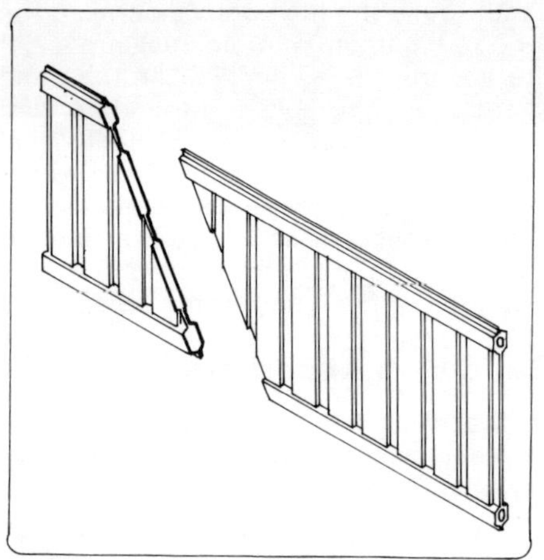

4.5 Central Heating Radiator Type.

contact with the whole of the absorber surface. Roll-bond aluminium absorbers and some types of central heating radiators used as absorbers fall into this category which thermally behave partly as sandwich types, and partly as tube-and-sheet types.

A critical design characteristic in sandwich absorbers is their water capacity. As explained in chapter 3, a high water content will mean a high thermal capacity which in turn results in a slow reacting solar collector. Ideally a solar absorber should have a water capacity of less than 0.4 pints /sq.ft. (2.5 litres/sq. metre) of absorbing surface.

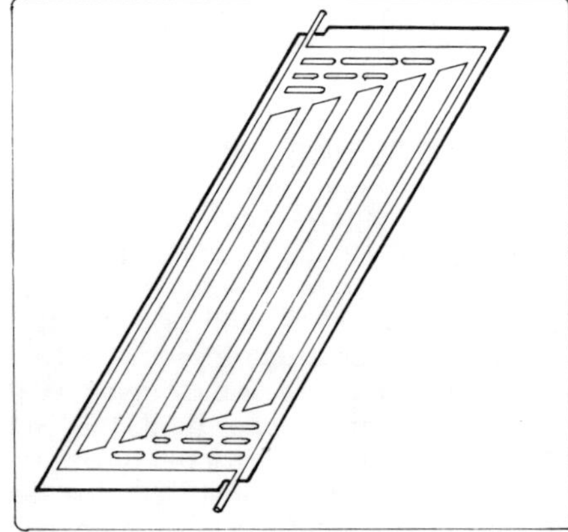

4.6 Roll-Bond Aluminium Plate.

From Sandwich to Breadbox

A different approach is to make the capacity of the sandwich very large so that it becomes a solar collector and storage vessel combined. This concept is inherently less efficient and suitable only where the hours of sunshine are long and uninterrupted. In their simplest form, such collectors could be simply a plastic covered trough of water. A more complex design currently gaining popularity in places like New Mexico is the breadbox solar water heater—a black painted hot water cylinder encased in an insulated box. The better designs feature an insulated lid which can reflect sunlight into the box by day, and hold the heat in at night.

Conduction through the Casing

If the case, or any part of it is metallic, it should not be allowed to come in direct contact with the absorber plate or its connecting pipework. This would form a 'heat bridge' enabling collected energy to leak away. Heat loss by conduction is reduced by insulating the case. The insulant must be able to withstand temperatures which may occasionally exceed 365°F (185°C.)

This rules out the use of plastic foam insulants such as styrofoam. Ordinary home insulation batts can withstand very high temperatures but they often contain a binder which vaporizes at about 300°F (150°). If they are used, they should be separated from the absorber by a layer of 1" thick rigid fibreglass duct liner.

Convection Currents

Convection currents around the absorber plate can be reduced by ensuring that the air gap between the plate and the cover is small—no more than 1 inch (25 mm).

The cover itself of course greatly reduces heat losses simply by shielding the absorber from the cooling effect of wind. Convection losses can be further decreased by ensuring that the insulation on the underside is pressed up against the plate so as to prevent any air movement on

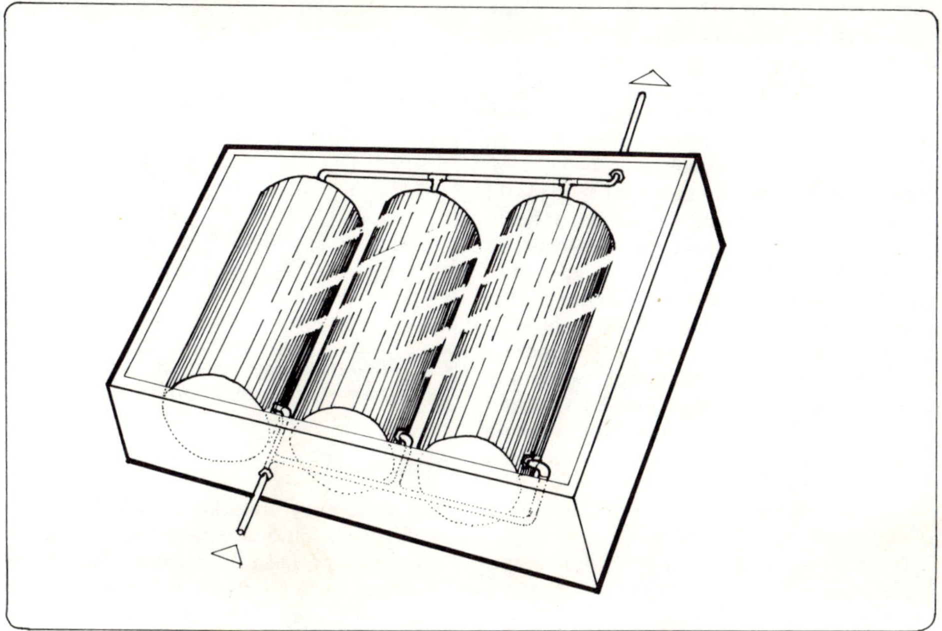

4.7 Breadbox solar water heater

the underside. Finally the case should be so designed as to have no air gaps which would allow through circulation of cold air. In practice though it is usually necessary, except in a few factory glazed and sealed collectors, to provide small vent and drain holes for the removal of condensation.

Re-Radiation Losses

The hotter a black body becomes, the more energy it will radiate. There are only two simple methods of reducing the radiative losses; through exploitation of the greenhouse effect and keeping operating temperatures to their lowest useful value. These have been described in the previous chapter.

High Efficiency Collectors

Various sophisticated techniques have been adopted to further decrease the heat losses from solar absorbers. These have in common a reliance upon industrial production methods difficult to reproduce in a home workshop, and considerable increases in cost. Bearing in mind the latter, it is worth noting that their effectiveness is most noticeable when operating at temperatures higher than is generally necessary for domestic water heating.

Honeycombs

Honeycombs of transparent tubes have been mounted perpendicularly over the absorber plate. This suppresses convection and, if the diameter of the tube is small in proportion to its height, it will reduce re-radiation losses. Care must be taken that the honeycomb is not in contact with the absorber and the cover, but rather suspended between them. Otherwise conduction losses, and perhaps re-radiation, will increase.

Selective Surfaces

Selective surfaces are extremely thin coatings electro-plated on to the absorber plate. They have the distinction that whilst absorbing a large proportion of solar radiation (not so much though, as a good matt black paint) the proportion of heat they re-radiate is very low. Their performance is often characterized by the ratio absorptivity:emissivity. An ordinary matt black surface might have a ratio of 90:90 whereas a good selective surface could have a ratio of 85:10. The emission factor is much lower and less heat is therefore radiated. This characteristic is very useful when operating at high temperatures. The selective surface is achieved by first applying a shiny coating such as bright nickel. This is then covered by a black absorbing layer such as black nickel. The thickness of this layer must be controlled so as to be only a few microns thick, i.e. less than the wavelength of the re-radiated heat waves.

The chemical structure of selective surfaces is such that they tend to be unstable in combination with copper and aluminium. There is therefore some doubt over just how many years they will continue to perform selectively at the values quoted for a new coating. For this reason there may be increasing interest in selectively coated glass. Coatings of tin oxide and indium oxide on the inner surface of a glass cover give results similar to that of a selective coating on the absorber. Such coatings have been used in other applications for many years without deterioration.

Selective coatings can usually be recognised by the Newton's rings, or rainbow coloured hue they reflect.

Evacuated Collectors

If the surroundings of the absorber plate can be partially evacuated, the convective and conductive losses will be greatly reduced. The absorber will retain more of its heat, just like coffee in a thermos flask. Evacuated glass tubes with a selectively coated waterway running through their centre are the most efficient solar collectors available.

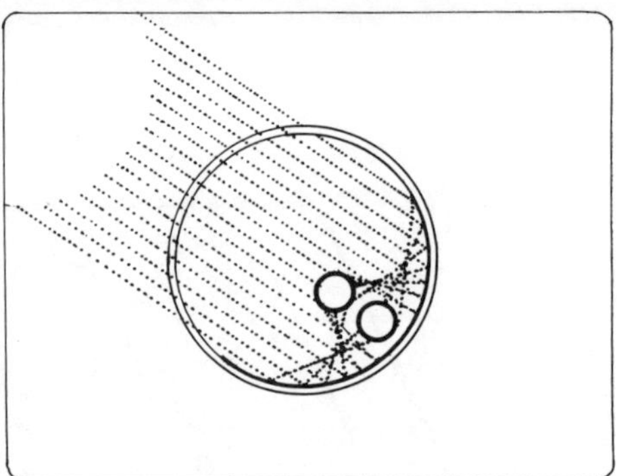

4.8 Evacuated tube collector, showing concentration of solar radiation on water pipes.

Applications

High efficiency collectors will prove most useful in situations where high temperatures are required, e.g. solar cooling appliances, certain industrial processes, and any solar installation required to produce mechanical movement such as solar powered pumps. They will also be of use where the system is required to work primarily in difficult conditions e.g. space heating systems whose main load comes in the winter. Finally, a point which has not been given so much attention, because of their higher efficiency a much smaller area of these collectors is required when compared to conventional collectors—as little as half the standard recommended area in some cases. This means that they will prove useful in retrofitting i.e. the application of solar system to existing buildings. When dealing with buildings which are already standing, overshadowing and awkwardly shaped roofs sometimes restrict the area available for mounting collectors.

Life Expectancy

Having paid attention to all the physical aspects affecting the collector's efficiency, there remains a less obvious quality which a good solar system must possess—longevity. Even a home made system will require several years to pay back in fuel savings the value of the materials expended in its construction. Unless its life exceeds the duration of this payback period, the resources devoted to its installation could have been better used.

Defects in a solar collector can come from within or without; from the corrosive effect of the circulating fluid acting on the absorber plate, or from weathering of the case and glazing.

Internal Corrosion

The combined presence of air and water always spells danger for thin sheets of metal. Collectors in drain-down systems and direct systems are therefore most susceptible to attack and only strongly resistant materials such as copper and stainless steel should be used for the waterways in such systems. Pressed steel absorbers should only be used in indirect systems (see p. 35) where the quantity of air dissolved in the circulating water is limited and cannot be replenished.

A major cause of corrosion is mixing of metals in the solar circuit. Different metals have different electro-chemical charges and because of this, when two different metals are immersed in water, there will be an increased tendency for one of them to dissolve. This is particularly noticeable where water flows from a copper pipe into aluminium or galvanised iron.

Manufacturers of aluminium absorber plates stress that they should only be used in indirect systems with anti-corrosive additives in the water and that they should be mounted so as to be electrically isolated from any other metallic elements.Furthermore they should never be connected directly to copper pipes. A rod of magnesium is sometimes used, suspended as a sacfrificial anode in the circulating fluid to 'use up' the corrosive ingredients in the water. It has been suggested the corrosion problems encountered with aluminium can be overcome by using oil instead of water as the circulating fluid which transfers heat from the collector to the store. Even with carefully selected oils however corrosion has occurred due to vapour becoming dissolved in the oil. Unless all the outlined precautions can be reliably maintained for the life of the installation, absorbers with aluminium waterways should be avoided.

Aluminium however can be useful when used as an absorbing sheet bonded to copper or stainless steel pipes in a tube-and-sheet absorber. In such designs, the joint between the two metals must be protectively coated in case moisture, condensation for example, should ever seep between them.

The plastics used in solar absorber construction are usually resistant to the corrosive effects of water. Hot water however can cause softening of most plastics and in some cases it can also leach out the stabilisers in the plastic. The ultra-violet component of solar radiation will also cause deterioration of many plastics. It is likely that suitable plastic absorbers will be developed in the near future, but for the present, it is safest to use them only in situations where the operating temperatures will be low, as in swimming pool water heating. One exception to this is the elastomer E.P.D.M. This synthetic rubber has been used by Bio–Energy Systems to produce their Solaroll absorber — a low cost and durable alternative to metal designs.

External Weathering

The casing which supports and protects the absorber, and keeps the insulation dry, obviously has considerable influence in determining the life of the collector. Aluminium boxes with welded corners appear to be the most durable casings available. Care should be taken during their installation that they will not receive rainwater run-off from any copper pipes. Similarly, any connecting pipes entering the case should be physically separated from the aluminium by the use of sleeves made from a durable material such as neoprene. Cases moulded from fiberglass also provide reliable protection for the absorber. If the casing has been manufactured from a less durable material such as wood, it will have to have regular attention to maintain its protective coating. Resinous timbers should be avoided as there is a danger that the resin will evaporate at high temperatures and condense on the underside of the glass thus reducing its transmissivity.

A weak point in many collectors is the joint between the glazing and the case. Conventional putty joints for example will have a very short life under direct sun and high temperatures if its surface is not regularly painted. Modern materials such as non-setting compounds, synthetic rubber, silicone sealants and pre-formed butyl strips are preferable alternatives, having life expectancies of 10–20 years.

Conclusion

Solar collectors for heating water have been around for over a century. But in the era of cheap fuel they were ignored. The unfamiliarity of solar collectors has led to some confusion regarding their function, and to mystification of their operation on the part of some salesmen. They are in fact quite simple heat exchangers, absorbing solar radiation and transferring it to water. If they are constructed with regard to the points laid out in this chapter there is no reason that they should not carry out this function satisfactorily for many, many years.

5

CIRCULATION

It is by circulating water through the solar collector that the absorbed energy is brought to the storage tank, or directly to the point where it will be used. Water can be moved round the system in one of two ways: gravity circulation, also known as thermosyphoning, and forced circulation achieved by means of a small pump or circulator.

As explained in chapter 7, the decision to use or not to use a pump is one of the first steps in planning a solar system. A pump will give much more freedom when it comes to locating the system components. Indeed, in some houses it will be impossible to circulate without a pump, because of the need to locate the storage tank above the collectors when relying on gravity circulation. A pumped system will also collect more energy than a comparable thermosyphoning system. This is because it will achieve higher flow rates which will mean lower operating temperatures in the collector and, as is explained in Chapter 3, this will lead to higher efficiency. Another way of explaining the improved performance achieved with a pump is that it enables one to maintain a higher rate of heat removal from the collector, thereby allowing the collector to absorb heat at a higher rate.

Cost Penalties with Pumped Systems

The disadvantages of opting for pumped circulation are increased costs and the introduction of reliance on an electrical supply, without which the system will not operate. A suitable pump will cost about \$100, the annual electrical bill will be about \$10 and automatic pump switch controllers sell for \$30—\$100. Against this can be set the savings accrued by using the smaller diameter piping which is suitable for pumped systems.

Thermosyphon Circulation

The need for an electricity supply is unlikely to be a problem in most cases. There is however an undeniable attraction about the independence and self regulating nature of the thermosyphon systems. In these, the storage is located above the collectors so that as water in the collectors becomes heated, its tendency to rise in a convection current, brings it up to the storage tank, at the same time drawing cold water from the bottom of the storage tank into the bottom of the collectors. The stronger the radiation arriving on the collector, the faster the circulation. And, of course, when there is no energy to collect, circulation will cease automatically.

The principle of thermosyphonage is very simple. To ensure that it works well in practice, some care must be taken in pipe layout, pipe sizing, storage location and the collector connection pattern. All these factors affect the resistance to flow through the circuit and this must be minimized.

Pipe Layout for Thermosyphon Circulation

Resistance to flow through pipes can be reduced by utilising larger diameter pipes, and minimizing the length of pipe and the number of bends and fittings. The convective current of heated water can also be facilitated by ensuring that, between the collectors and the storage tank, there is a continuous rising gradient in the pipework.

Similarly, there should be a continuous fall in the piping connecting the bottom of the storage tank to the bottom of the collectors.

No horizontal piping should be installed in a thermosyphoning loop. A minimum slope of 1 in 40 is a good rule.

In the horizontal plane, the collectors and tank should be as close as possible to each other, in order that the length of pipework be small. In the vertical plane, increasing the height of the tank above the collectors by up to about 13 feet (4 metres) can improve the thermosyphonic flow. Greater separation will cause a drop in circulation rate. The advantages of a large vertical separation between collectors and storage might be outweighed however if it also meant longer runs in the distribution pipework between the storage tank and the points of use, as this would result in increased heat losses.

In all cases, the storage tank should be at least 2 feet (600 mm) above the top of the collectors to ensure that there will be no back syphonage on cold nights. If this occurred, warm water from the storage tank might be drawn into the collectors where it would be cooled by radiating its heat to the night sky.

Pipe Diameter in Thermosyphon Circulation Sytems

Although I have produced tanks full of hot water on very sunny days using ½ inch (15 mm) diameter pipe, I would not recommend anyone to use pipe smaller than 1 inch (28 mm) for a domestic installation.

Solar Collectors for Thermosyphon Systems

As shown in the last chapter, there is a considerable variety in the construction and configuration of solar collectors. Some are better than others for facilitating a thermosyphonic flow.

The open trough type, trickle collector and coil configurations cannot be used in a gravity circulation system as they will not function without a pump.

Tube on strip collectors where the tube zig-zags across the plate in a serpentine configuration can offer considerable resistance to water flow. The best ones are those which use at least ¾ inch (22 mm) pipe and where there is an obvious gradient in the 'horizontal' pipe runs.

Sandwich type collectors, such as radiator panels, will offer comparatively little resistance to flow. In the case of radiators however, their water capacity and hence their thermal mass, is high. As explained in chapter 3, this will mean that the collector will not effectively respond to short outbursts of sun. If this type of collector is used, it should be mounted so that any ribs or reinforcement which form channels inside the absorber, run vertically so as not to obstruct the upward flow of warm water. They should also be mounted at a slight tilt so that the horizontal headers have a slight gradient.

Tube on strip collectors which have a grid layout of parallel tubes between two horizontal header pipes are the best choice

5.1 Thermosyphon (Gravity Circulation)

1. Solar collector
2. Pipe carrying hot water from collector to storage
3. Storage tank
4. Cold water supply to storage tank
5. Pipe carrying hot water to faucet
6. Pipe carrying water from storage to the bottom of the collector
7. Header pipe which must be installed with a small gradient to facilitate circulation

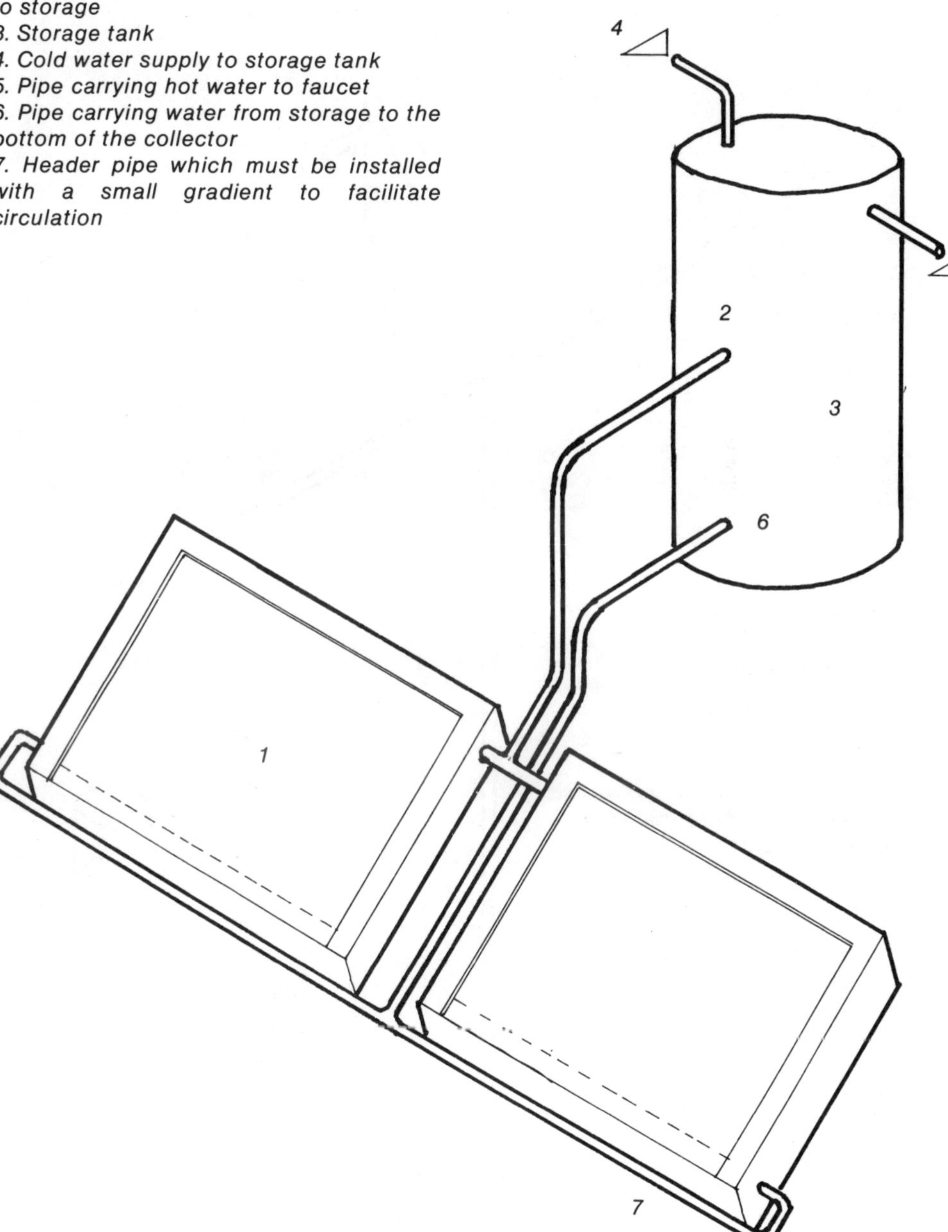

5.2 Forced Circulation (pumped)

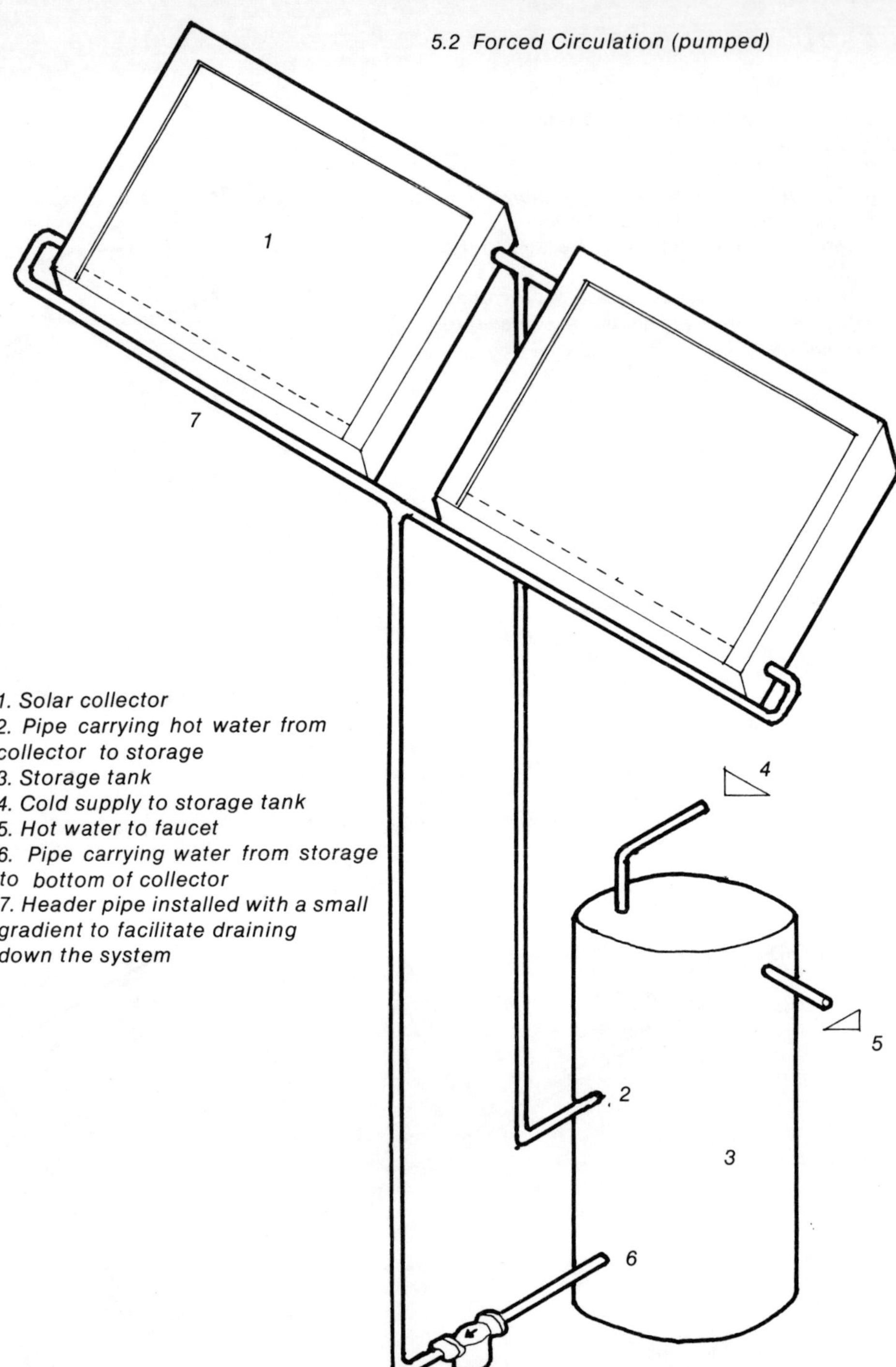

1. Solar collector
2. Pipe carrying hot water from collector to storage
3. Storage tank
4. Cold supply to storage tank
5. Hot water to faucet
6. Pipe carrying water from storage to bottom of collector
7. Header pipe installed with a small gradient to facilitate draining down the system

for a thermosyphon system.

Because of the slower reaction of thermosyphoning systems, and their higher operating temperatures, double glazing and increased insulation may be advantageous in the collectors.

Choice of Pump

If circulation is to be forced, a suitable pump must be selected. A centrifugal circulator is the best choice. It should be of the canned rotor, or magnetic driven type. The alternative, an external motor with a drive shaft connection, is more likely to give problems with leaks, especially around the shaft seal. It is important that the pump should be designed for operation at high temperatures.

Only a small pump is required—about 1/20 to 1/12 h.p. will be about right. The choice of pump rating can be simplified by choosing a variable head pump. These are now produced by the major manufacturers at the same price as the single rated pumps. The pump which I am most familiar with is a Grundfoss model which also has a two way speed selector switch. When installed the pump should be set at its lowest rating and speed. If the flow is too slow, as indicated by high temperature differences between the collector inlet and outlet the speed can be increased. If this is still inadequate, the rating selector can be changed.

Pumps in Drain down Systems

A slightly stronger pump is needed if the collectors are to be drained down when there is insufficient energy to warrant circulation. A submersible pump inside an open top storage tank is appropriate. Most submersibles are designed for low temperature applications like fountains, and should be avoided. High temperature models are much more expensive. Other pumps can be used but care must be taken in locating them well below the level of water in the storage tank. Most pumps are not self priming and if there is not water in the pipework on each side of the pump when it is turned on, it is likely to burn out its motor.

Pumps in Direct Systems

Direct systems suffer much more from the effects of corrosion since fresh aerated water is continually being introduced.Only bronze pumps should be used in direct systems,as these are less liable to corrode.

Control Switch for Pump

As sure as night follows day, there will not always be enough energy being collected to warrant running the pump continuously. A control mechanism is required.

In sunny climates,a manual switch or a simple time switch will suffice,turning the pump on in the morning,and off in the evening or in mid-afternoon. In cloudy countries where the pattern of sunshine is less predictable,a system left running all day will frequently be radiating heat to the sky instead of vice versa. Unfortunately it will not improve the weather, and it certainly will not decrease the fuel bills.

The pump could be controlled by a radiation sensitive device such as a solar cell which would switch the pump on whenever solar radiation exceeded a preselected level, say 50 Btu/hr/ft^2 (150 w/m^2). This could also be done by an ordinary room thermostat mounted on the back of a small sheet of blackened aluminium set inside an insulated glazed box, a mini solar collector, mounted beside the real ones. The thermostat should be given a low temperature setting, but remember that the sheet of aluminium will heat up much quicker than the water filled panels. The thermostat should be kept accessible so that adjustments can be made to find the best setting for any particular situation.

Such radiation sensitive controls can be made without spending a lot of money. There will however be occasions when the system will be throwing heat away, e.g. after a bright sunny morning the storage tank might be full of very hot water. If the afternoon is only moderately bright, the radiation sensor is still likely to keep the pump running and hence lower the tank temperature.

The most accurate type of pump controller for solar systems is the one known as a differential thermostatic controller, otherwise called a 'black-box'. The box is fed by four sets of wires; one brings power from the electricity supply, a second takes power to the pump; the remaining two are connections to temperature sensors, one attached to the storage tank, the other attached to the collector outlet. Inside the box, a simple electronic circuit compares

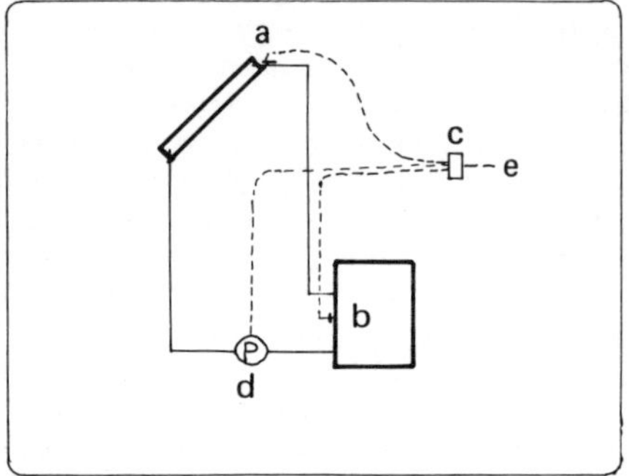

5.3 Diagram of Pump Controller Layout. a) temperature sensor mounted on solar collector; b) sensor on storage tank; c) 'black-box' control; d) pump; e) electricity supply.

the two temperature readings. Only when the collector is warmer than the water in the storage tank will the pump be switched on.

Controller Improvements

The electronic circuit is sometimes made a little more sophisticated by adding a selector switch which allows the installer to choose by how many degrees the collector should be warmer than the tank before starting circulation. The usual setting is about 6°F (3°C). This difference is to ensure that the energy collected is greater than the amount of energy required to run the pump.

A problem frequently encountered with control systems is that known as 'cycling', or 'hunting'. When this happens, the pump continues to switch on and off in very short time cycles. The problem here is that water in the collector becomes heated and the differential controller switches the pump on. This brings an immediate rush of fresh cold water into the collector which cools it down thus causing the controller to turn the pump off. This becomes a serious problem in systems with long pipe runs. Warm water from the collectors would be forced out of the collectors but may never reach the tank before the pump is switched off. This would mean that the heated water would lay in the flow pipe and would perhaps have cooled down before the collectors became hot enough to cause the pump to switch on again.

Some controllers counter this problem by including a time lapse device which will prevent the controller from turning the pump off until a predetermined interval after the collector is cooled. The time interval required is equal to the time taken for water to flow from the collector inlet to the tank inlet. This period can be measured by leaving the pump switched off on a sunny morning so that the pipework is cold and the water in the collector very hot. With one hand on a stop watch and the other on the tank inlet, one can measure the time between switching on the pump manually and feeling the pipe heating up. This indicates the time water takes to flow to the tank from the collector outlet. After a few seconds the temperature will fall again, indicating that all the water that was laying in the collectors has been pumped through to the tank. This is the time interval which should be registered on the controller.

A further sophistication in some controllers is a temperature related time lag which will allow the pump to continue running for up to 3 minutes if the collector is only 1°F (½°C) cooler than the tank. The greater the cooling effect of the solar collectors, the sooner the pump will be cut off. Some pump/controller units can vary the rate of circulation in proportion to the collector's operating temperature. Pumping faster as the collector becomes hotter is a way of increasing its efficiency.

Fixing the Thermal Sensors

The thermal sensors, usually thermistors, upon which the differential temperature controller relies, must have a good thermal contact with the collector outlet and the tank. They can be fixed with an epoxy resin, tucked into a tight fitting copper sleeve which has been previously soldered onto the pipe, or strapped on firmly with a non-setting thermal paste to aid thermal contact. They should both be well insulated after fixing. The most accurate responses come from sensors immersed in the water.

The collector sensor should be situated immediately outside, or even inside the collector casing. The storage tank sensor should be situated midway between the inlet and outlet leading to the collectors.

Freeze Protection

Due to the exposed position of solar collectors, there is a danger that the circulating fluid might freeze during cold nights. The danger is greatest on nights with clear skies as the collector temperature can drop below the surrounding air temperature due to radiation of heat into space.

As water approaches its freezing point, it expands in volume. This expansion can fracture the metal waterways causing them to leak when the liquid defrosts. Steps must be taken therefore to prevent freezing of the fluid in solar collectors. There are five possible solutions to this:

1) The use of a non-freezing heat transfer fluid.
2) Draindown of collectors.
3) Heating the collectors, either by circulàting warm water from the tank or by an electrical resistance input.
4) Insertion of compressible gas-filled tubes inside the waterways to accept the expansion experienced on freezing.
5) Insulated covers.

Non-Freezing Fluids

Several specially formulated non-freezing fluids are now available for solar heating systems. The three types in use are water/anti-freeze solutions, hydrocarbons and silicone fluids. All have the primary disadvantage that their use requires the separation of the fluid circulating through the collectors from the water in the storage tank which will eventually be used in the house. This separation is achieved by installing a heat exchanger to close the collector circulation loop. Systems incorporating heat exchangers in this way are called *indirect systems*. Heat exchangers can never be 100% efficient so indirect systems start off at a disadvantage to direct systems.

Water/Anti-Freeze

The most common anti-freeze additive is ethylene glycol, often used in automobile cooling systems. It is however a toxic substance and if it were to leak through the heat exchanger into the supply water, it could cause poisoning where water from the hot water system is swallowed in large quantities. Because of this danger, most water authorities insist on the use of a double-walled heat exchanger. Propylene glycol is sometimes accepted as a non-toxic alternative to ethylene glycol.

Both of these anti-freeze solutions however are subject to decomposition into glycolic acid at high temperatures. This is highly corrosive and inhibitors must therefore be added to water/glycol mixtures. An annual check on the fluid's ph value, its freezing point and the concentration of inhibitors is advisable. Otherwise it is advisable to replace the fluid every 2 years unless your supplier can guarantee a longer life.

How much Anti-freeze?

The quantity of anti-freeze required depends upon the water capacity of the primary circuit pipework and the collectors, and upon the physical properties of the solution in reducing the freezing point. A typical domestic system will hold less than 8 gallons (36 litres) in the collector loop.

If you wish to calculate the capacity of your own system, you will have to check the trade literature for water content of the particular type of collectors you are using, and measure the length of pipework, not forgetting the heat exchanger coil. Knowing the length of pipework, the following table will help you calculate its water capacity.

Nominal Tube Diameter	Water Contents
½ inch	0.012 gals/ft
¾ inch	0.026 gals/ft
1 inch	0.043 gals/ft

To determine what percentage of the total capacity should be made up with anti-freeze, first find out the lowest winter temperature ever recorded in your locality, then select from the following table a percentage. The following table will then help you select the percentage of ethylene glycol which would be necessary to see you through the coldest night. Note that propylene glycol has a slightly higher freezing point. A 50% mix of propylene glycol would freeze at -28°F.

Ethylene Glycol Solution (% by volume)	Approximate Freezing Point
20%	+16°F
30%	+ 4°F
33⅓%	0°F
40%	−12°F
50%	−34°F

Hydrocarbons

Hydrocarbons, particularly mineral oils, are popular in systems using aluminium collectors. Their use requires greater pumping power as their specific heat is almost half that of water meaning that you need to circulate at almost double the speed to extract equal amounts of heat from oil-filled than water-filled collectors. The pump's work is further increased by the fact that the viscosity of the oils is higher than that of water—in other words, it is more sluggish in its journey through the collectors. At high temperatures there is again the danger of the fluid decomposing into acids and sludge. Regular checks on the composition of heat transfer oils are therefore advisable, but replacement will not be so frequent as with water/glycol solutions. Oils should be used only in sealed systems due to their inflammability and care is required in the selection of compatible sealants for pipe joints and any rubber components. Many standard pumps and valves designed for use with water may not be compatible with heat transfer oils.

Silicone Fluids

Silicone heat transfer fluids are about 5 times more expensive than the alternatives. This is more than compensated for by the fact that they eliminate the necessity for annual checks on the fluid's composition. They will not break down, freeze, boil, burn, damage building materials, nor corrode metals. Their life expectancy is as long as the collectors themselves. Unfortunately they have a higher viscosity than either water or heat transfer oils, and also a lower specific heat. Thus greater pumping power is required and the heat exchanger should be increased in surface area. Both silicone and mineral oils are able to escape through joints which would hold water. Special attention therefore is advisable when installing the pipework and soldered joints are recommended.

Draindown of Collectors

Draindown systems require more pumping power and are more liable to corrosion due to the continual introduction of air into the wet inner surfaces of the collectors. Special attention is also needed to prevent air blockages occurring. Copper and stainless steel should be able to withstand corrosion attacks for many years, but a more difficult problem is the formation of scale in regions with hard water supplies. This deposit on the inside of the collectors reduces the transfer of heat between the absorber and the circulating water. It would have to be chemically removed every few years.

Heating Collectors at Night

A simple thermostat in the collectors could start circulation of warm water from the storage tank, or switch on electrical resistance heaters inside the panels whenever their temperature approached freezing point. This at first seems a terrible waste of energy but in some situations the energy gains achieved by operating without a heat exchanger can be greater than the energy expended on heating on cold nights.

Compressible Tubes

The insertion of gas-filled compressible tubes inside the collector's waterways is an attractive method of dealing with the freezing problem. There are of course practical difficulties and as yet the technique has not been widely applied.

Insulated Covers

Covers which fit over the face of the collectors and prevent freezing at night are probably the simplest and cheapest solution. Unfortunately it depends upon the frailty of human memory unless automatic blinds are introduced and that would be very expensive. The risk of damage if the collectors were left uncovered is probably too great to make covers a practical solution in all but a few cases.

6

STORAGE

The heat store is the root of the solar system. Without it, hot water would only be available when the sun is actually shining. A storage unit can enable the system to collect heat whenever it is available, and to deliver it when it is required. In the real world however, it would be hopelessly uneconomic in most climates to install a solar storage unit large enough to meet all the year-round hot water demands of the typical household. In practice therefore the storage in a solar hot water system is designed to hold one or two days supply of hot water and an auxiliary heating system is included to meet the load whenever there is a longer run of days with insufficient sunshine.

In water heating solar systems, the heat is stored, sensibly enough, in water; either water which will eventually be piped to the faucet, or water which is kept in the storage tank and passes its heat to the faucet water via a heat exchanger.

This chapter looks at the types and sizes of tanks suitable for storing solar heated water, and the heat exchangers which might be used to transfer heat into, or out of, the storage vessel.

Choice of Tank

The best choice for a storage tank will be the kind of steel cylinders currently used in conventional hot water storage systems. These should be protected from corrosion by an internal layer of stone (concrete), or glass. Stone-lined is preferable as there have been reports about problems with glass-lined tanks, particularly in cases where they have been fitted with copper heat exchangers.

All-copper cylinders have the combined advantages of superior corrosion resistance, and lightness. The weight may be an important consideration if you are installing the system yourself. Copper cylinders are selected almost automatically in countries like England, where regulations insist on low-pressure hot water systems which means that thin-walled, relatively cheap tanks are acceptable. Unfortunately these are not widely available in the USA. Tanks holding more than 120 gallons, and subjected to high pressure by direct connection to mains water supply, must carry an ASME label certifying that fabrication meets safety standards.

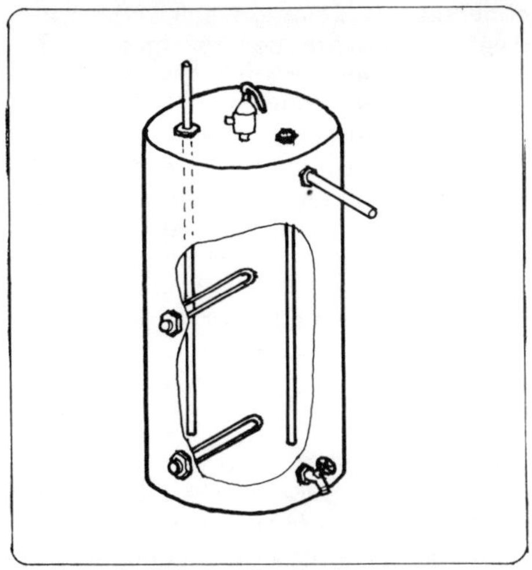

6.1 A typical hot water tank

Suitably located inlets and outlets are important features to look for when selecting a tank. Five threaded connections are desirable: two at the top, for a pressure-temperature release valve, and a hot service water draw-off; two at the bottom for cold water inlet, and the outlet to the collectors; and an inlet for heated water from the collectors about ⅓ or ⅔ up the tank. Note however that in some tanks the cold water supply enters at the top of the tank and is carried to the bottom via an internal dip tube.

The storage tanks described above will represent a major proportion of the total system cost. There is variation in prices, so it is worth shopping around. Various alternatives to conventional cylinders have been suggested but when considering these, bear in mind that you want the system to operate without trouble for many years. Remember also that leaks in storage tanks can be pretty devastating due to the amount of water they contain. This is not so critical if you are siting the tank in your back yard, but if it is to sit in or above your living area, a cheap tank may prove to be a false economy.

Open Top Tanks

One alternative you might like to consider is the use of open top polypropylene or fiberglass tanks. As they are open to the atmosphere, pressure ratings present no problems. But care is needed in selecting tanks which can withstand boiling temperatures. Apart from their lower cost, open top tanks make possible further savings because it is possible to install in them home-made heat exchangers—which can be simply a coil of pipe. This would be extremely difficult to install inside a closed cylinder. Open top tanks however suffer greater heat losses than cylinders because of evaporation which occurs across the

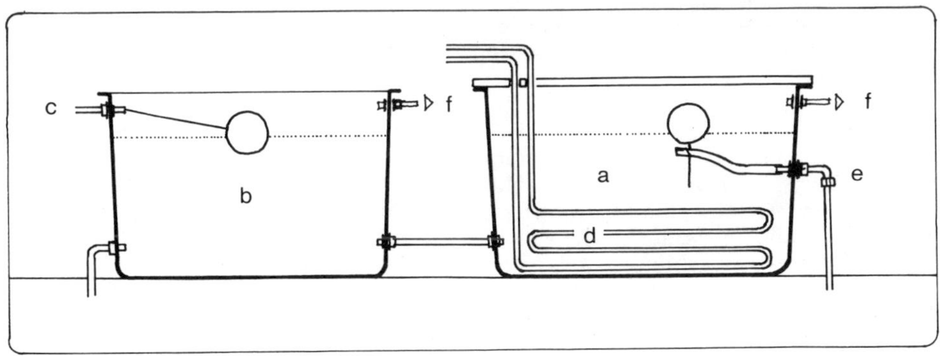

6.2 a) Open top tank level with b) cold storage/feed tank. c) Water supply via ball valve. d) Heat exchanger. e) Warm water draw-off via flexible plastic tube connected to float to ensure that draw-off is taken from the uppermost layer of water. f) Overflow pipes.

air/water surface. Furthermore the drop in the water level which occurs during draw-offs tend to draw in cold outside air. This absorbs heat which is then expelled as the tank refills. This will occur even when a tight fitting lid is used. The solution to these heat loss problems is to use a floating layer of insulation. Polypropylene balls are manufactured for similar industrial applications.

A more easily obtainable material would be polystyrene granules. This material might melt however if the water temperature exceeded 149°F (65°C). The insulating balls should be of a diameter larger than the ¾ inch (22 mm) draw-off pipe to prevent blockages occurring. A cover is still required to keep dust out of the tank. The best choice are tight-fitting plastic covers, designed to drip any water which condenses on their underside, into the tank and not over the side.

Obviously an open top tank cannot be filled directly by the supply pipe — it would continually overflow. What is needed is a float valve. This is fitted above the desired water level in the tank, and it automatically closes off the supply whenever water in the tank reaches the pre-set level. This is achieved by the lever action of an arm projecting from the valve. The arm has a hollow ball fixed to its end and this ball is forced upwards by the rising water level. Adjustments in the cut-off level can be made crudely by bending the ball's arm in order to alter the height at which the ball exerts enough pressure to close the valve. It is the same device you can find in most toilet tanks.

The float valve could be fitted directly on the solar storage tank, but it would require a down pipe to be suspended below it in order to carry incoming coldwater to the bottom of the tank rather than allowing it to mix with, and cool the hottest water which will gather at the tank's surface. In this case it would also be important to establish that the ball of the valve, which is usually made of plastic, can withstand the temperatures at the top of the tank. Some local codes may forbid direct connection of the supply via a float valve to the hot tank. If this is so, the supply will have to be fitted to an adjacent cold water storage tank whose water level will have to be the same as that desired in the solar store. Cold water can then be piped from this tank to the bottom of the solar store.

In all open top tanks; a ¾" overflow pipe must be fitted above the intended water level, in case the float valve develops a fault. When you buy tank connectors be sure to say what type of tank you are using so that you can get the correct gaskets for sealing the joint.

How Big

The size of the storage tank should be related to the amount of hot water used each day. Few of us know how many gallons we actually use, and the amount varies a great deal from one household to another. It depends to some extent on the number of occupants and to some extent on the appliances in use. An automatic dishwasher for example, might use 15 gallons compared with about 4 gallons for dishwashing

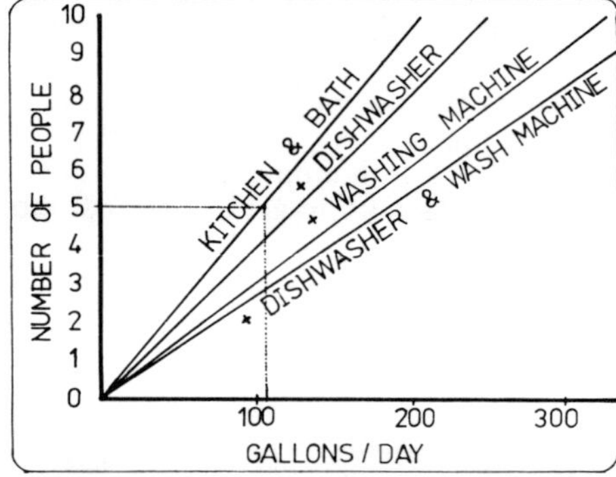

6.3 Hot Water Requirements (derived from Copper Development Assoc. publication). Locate on the vertical axis the number of people in the household. Draw a horizontal line through this point and across the four diagonal lines. These represent four classes of houses differentiated by the appliances they contain. Select one according to the appliances you own. Where it meets the horizontal line, drop another line to the gallons/day axis below. The point where your vertical line meets it will indicate your home's probable daily hot water requirement.

by hand. Washing machines are perhaps the biggest consumers, taking about 21 gallons per load. Showers, at about 10 gallons a dip, draw considerably less than baths which take 15 gallons, or more if you like to lay and wallow in it.

Figure 6.3 should help in arriving at a sensible estimate for your own home. Following the instructions given in the figure's caption it shows for example, that a 5 person household with a standard kitchen and bath uses a little under 110 gallons each day. Standard sizes for hot water tanks vary a great deal. In the example we are considering, we would probably select a tank with a capacity of about 125 gallons to serve as the solar storage tank.

Where To Put It

It must be remembered when installing storage units that a tank full of water is very heavy. Each cubic foot of water weighs 62 pounds. A 125 gallon tank will weigh over 7800 pounds when filled. That's why most solar tanks end up in the basement where the load can rest on bedrock. In some houses this will not be possible and indeed with certain systems which have been discussed, such as thermosyphoning types, a basement location will almost certainly be too low. If a tank is to be located on a suspended floor or in the roof space, try to locate it above a load bearing wall, and spread its weight on bearers. If you are at all uncertain about the ability of the structure to support the load, it is worth seeking professional advice.

6.4 Plastic open tank mounted on a flat even surface with weight spread on bearers

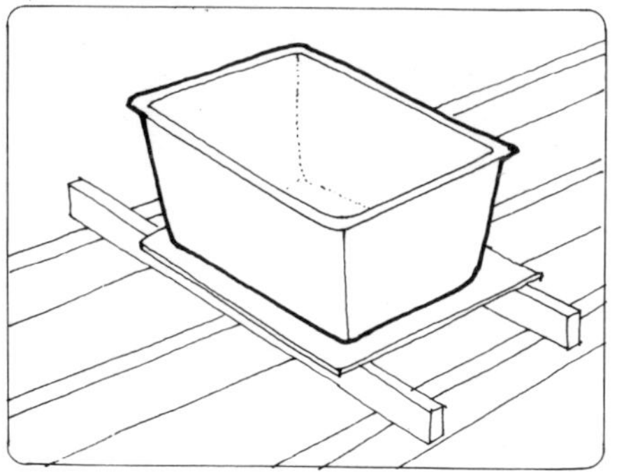

If a plastic tank is to be used, note that it must be supported across the whole of its base surface. A suitably sized sheet of ¾" plywood should be laid over the ceiling joists, or bearers, to support the plastic tank going into a roof space.

Heat Exchangers

In chapter 7 the choice of direct or indirect systems is discussed. In indirect systems, a heat exchanger must be connected to the storage tank. This serves to separate the water flowing through the collectors from the water which flows to the faucets inside the house. It must of course allow the heat absorbed by the water in the collectors to pass into the water in the storage tank.

No heat exchanger can transfer all the heat absorbed by the primary circuit into the secondary circuit. Normally only 60—90% is exchanged at each pass. A proportion of the solar energy collected will never reach the storage tank because of the inefficiency of heat exchangers. It is worthwhile therefore looking at the factors which affect their performance.

The amount of heat transferred across a heat exchanger depends upon four conditions:

(i) The temperature difference between the hot water flowing into the exchanger and the water to be heated in the tank.

(ii) The thermal conductivity of the heat exchanger i.e. the ease with which heat can flow through the walls of the heat exchanger. Metals generally have a high thermal conductivity and copper is particularly good.

(iii) The surface area of the exchanger. The larger the area of interface between the hot and cold water, the more heat can flow between them.

(iv) The water flow rate. The more water that flows through the exchanger the greater will be the quantity of heat that passes through it.

Location of Heat Exchanger

The heat exchanger should be in contact with the bottom of the solar storage tank where the coldest water settles. This is where the greatest temperature difference can be found, and hence where the exchanger will operate most effectively. This is extremely important in solar systems as the hot water coming from the solar collectors will on many occasions be only a few degrees above the temperature of the water in the tank.

6.5 Heat exchange coil in the bottom of a hot water cylinder

Heat Exchanger Material

The best material for constructing a heat exchanger is copper. Plastics are generally insulative and tend to resist the through flow of heat. A cheap solution which has been suggested by recycling enthusiasts is the use of heat exchangers salvaged from old refrigerators or automobile radiators. These may prove effective for a short period but will undoubtedly run into problems with corrosion after a few months.

Heat Exchanger Surface Area

To ensure that the heat exchanger will not reduce the overall solar contribution by more than 2 or 3%, the heat exchanger surface area should be at least 15% of the solar area. This means that if the exchanger is to be made of 1" tube, you should have 6" length in the exchanger for every 1 square foot of collector. Smaller diameter pipes have smaller surface areas and hence a larger length would be required to form a comparable heat exchanger; 8"/sq. foot with ¾" pipe, and 11"/sq. foot with ½" pipe. Where space is a problem, finned copper pipe, which has a larger surface area than regular pipe, can be used to form more compact exchangers. A range of different types are available from companies such as Spiral Tubing Corp.

Note that the above recommendations refer to simple exchangers made from coils or grids of pipe submerged in the storage tank. If you will be using a non-potable fluid in the solar collector loop, it is probable that your local code will require that you use a *double-walled* heat exchanger to minimise the possibility of contamination of the hot service water which is delivered to your kitchen. A double wall in a heat exchanger will obviously be more resistant to heat flow and therefore a larger surface area is desirable in order to counteract this problem.

Heat Exchanger Geometry

The geometric form of the heat exchanger will affect the rate of flow of the water passing through it. Sharp bends increase the resistance to flow, slow it down, and thus decrease the rate of heat transfer. Gradual 'slow' bends are therefore preferable to sharp elbows. This consideration

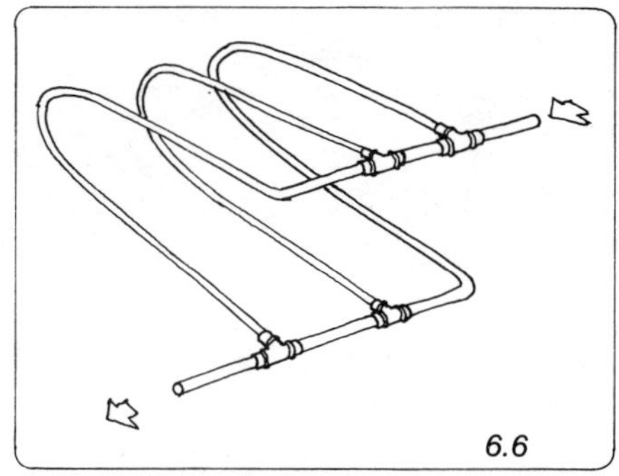

6.6

is of particular importance in thermosyphoning systems where the flow is already much slower. In such cases a grid of pipes, as shown in figure 6.6 will present less resistance to flow than a long length of pipe with many bends.

In pumped systems though, a much simpler exchanger can be made. Using a plumber's bending spring as explained in chapter 8, form an appropriate length of pipe into a serpentine, zig-zag arrangement. This should be dimensioned to fit into the bottom of the tank walls. It should also be supported inside the tank to raise it clear of the base, where it might become covered by sludge deposits. Bricks can serve as supports.

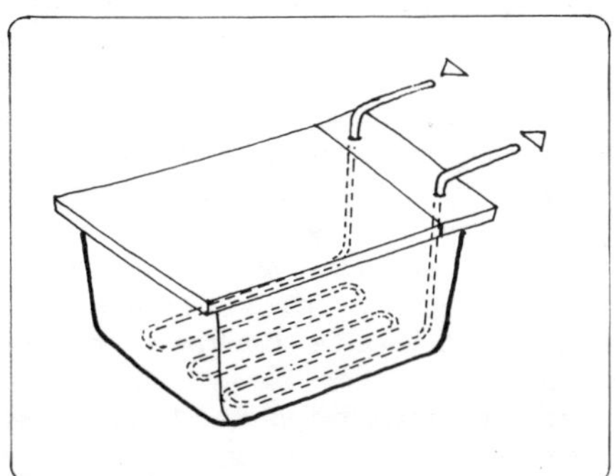

6.7 Serpentine heat exchanger for pumped systems

Soft annealed copper tube is best for home made exchangers as it is easily flexed into the desired shape. It is sold in coils which could simply be spaced apart and submerged in the tank. Use a length of scrap tube and some copper wire to fix a 1" (25 mm) spacing between the coils to encourage circulation around them. In pumped systems, where resistance to flow can easily be overcome, bends in the heat exchanger can be beneficial. They can increase the flow turbulence and hence the heat transfer.

Connecting the Exchanger in Open Top Tanks

With pumped systems the connecting pipes between solar collectors and heat exchangers can be simply dropped into the tank from above as in figure 6.6. It would be wise to insulate the outlet of the heat exchanger with some waterproof insulation where it passes vertically through the upper layer of water in the storage tank. Otherwise it will on occasion be taking heat away from the top of the tank.

In thermosyphoning systems, dropping pipes from the top of the tank into the heat exchanger forms additional resistance to flow and it is therefore better to connect the exchanger through holes drilled in the walls of the tank, as shown in figure 6.8.

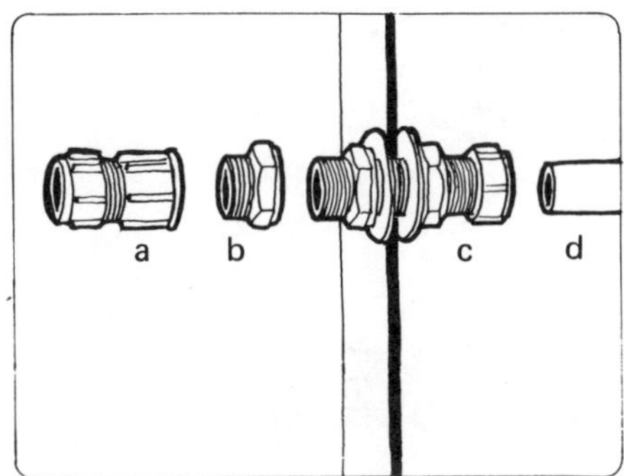

6.8 Heat Exchanger Connection through Tank Wall. a) straight couple; copper to female iron b) Reducing bush c) straight couple; copper to male thread with extended shank, two backnuts and washers d) copper heat exchanger.

Heat Exchangers in Closed Tanks

Hot water cylinders with built-in heat exchangers are commercially available, but you are advised to select one which is manufactured especially for the solar trade by companies such as Ford Products Corp. (stone-lined tanks) and A.O. Smith Corp. (glass-lined tanks). Such tanks should have extra large heat exchangers, possibly double-walled, and suitably located inlets and outlets.

External Heat Exchangers

If you are wanting to use a cylinder which

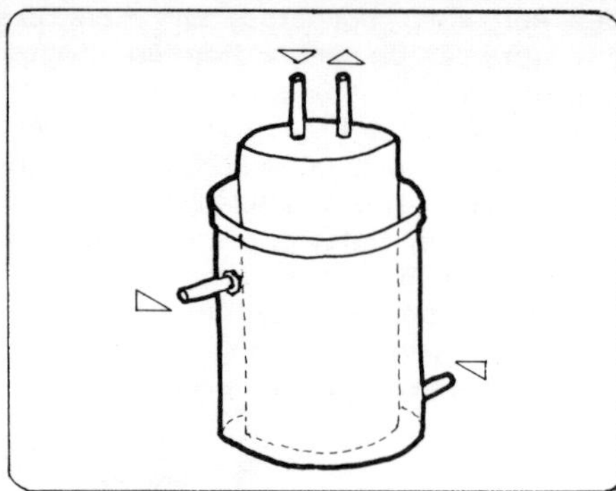

6.9 An open top tank used as a heat exchanger

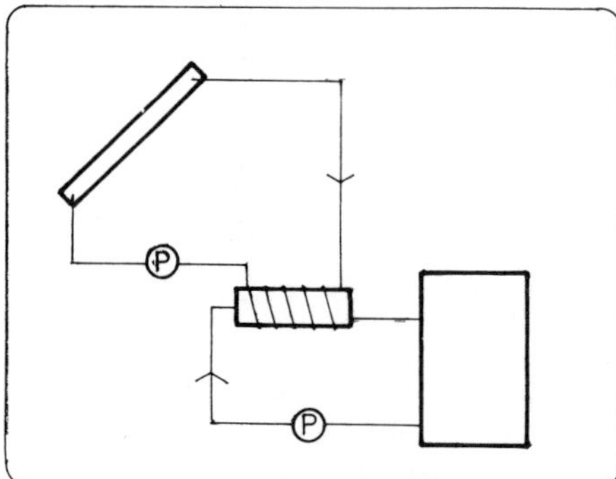

6.10 Two pump system with external heat exchanger

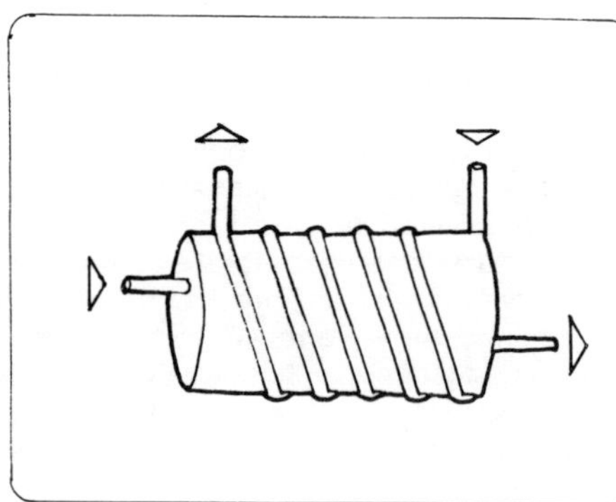

6.11 External heat exchanger

does not have an integral heat exchanger fitted, it is possible to fit an exchanger to the outside of the tank. If however your tank is lined to prevent corrosion, the lining will form an effective barrier to heat transfer. There are two ways of making an external heat exchanger. You can coil copper tubing around the bottom of the cylinder, or you can partially submerge the cylinder in an open top, cylindrical polypropylene tank whose internal diameter is 2 inches greater than the external diameter of the solar storage cylinder. The copper coil method relies upon good thermal contact between the tube and tank for successful heat transfer. The coils must therefore be soldered to the cylinder, or else clamped or wired into tight contact then wedged with a heat transfer cement such as Thermon, to increase the area of thermal contact. This method provides a double-walled exchanger.

The tank-in-a-tank method has been suggested by the Ecotope Group from Seattle as a simple method for forming a large-area heat exchanger which is particularly suited to thermosyphon systems because of the low resistance it presents to circulation. The system works best if you are using a copper tank as the solar storage tank. Ecotope point out that anti-corrosion additives will be necessary if you are using a steel cylinder and recommend a surface layer of oil to prevent evaporation losses from the outer tank. Notice that this tank-in-a-tank would be classified as a single-walled exchanger.

Separate Heat Exchangers

Code requirements for double walled heat exchangers can be met, and maintenance made easier, by using a separate exchange unit which is mounted outside the solar storage tank.

Its disadvantage is the increased initial and running costs, due to the extra pump which is required to circulate water between the tank and the exchanger. This pump, by the way, would have to be wired to the same temperature differential controller which runs the pump on the solar collector loop, and this may affect your choice of controller. Separate exchange units are available in both single and double-walled versions. In areas where the water is liable to deposit a mineral scale when heated, it is advisable to select a heat exchange unit which has a removable head to facilitate internal cleaning.

Insulation

Do not forget that even with the best tanks and heat exchangers, very little heat will be stored without the help of insulation. Tanks should have at least 6" (125 mm) of a good thermal insulating material wrapped around them. Use extra at the top where the tank will be hottest.

There are two ways to fit the insulation. You can use batts of fibreglass and tie these around the tank. Fixing them with duct tape and staples may be an easier method. Alternatively you can build a box around the tank and fill this with fiber-fill or styrofoam beads. The box can be made of wood or styrofoam boards. The latter will probably be more convenient as you have to cut several holes through the box to accommodate the pipework.

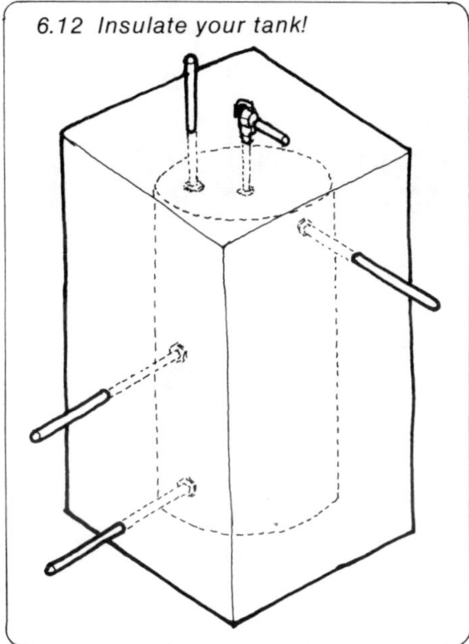

6.12 Insulate your tank!

Safety Valve

If you are using a closed cylinder as a heat store, remember that it will be operating at pressures greater than atmospheric pressure. Also, when it is heated its contents will tend to expand thus increasing the pressures within the tank.

Furthermore solar tanks can easily boil if they are small in relation to the collectors or if there is a long sunny period between draw-offs. It is essential therefore that a temperature and pressure relief valve be fitted in the top of such tanks. The valve should be connected to a copper tube which discharges over a drain whenever pressures exceed 125 p.s.i. or if temperatures exceed 210°F.

Conclusion

We have now examined all the major parts of a solar water heating system. In the following chapters, we shall be looking at how to put all the pieces together.

7

CHOOSING A SYSTEM

The solar collector, a storage unit and a means of circulating heat between them are the common elements we can find in any solar water heating system. There is a huge variety in the ways the parts can be put together, and how they connect to the conventional water heating system. This chapter has been planned to guide you through the maze of alternatives you might encounter when you come to choose a system layout.

There are four things you have to consider in coming to a choice:

1) cost and ease of installation,
2) available locations for collectors and tank,
3) freezing protection (where necessary),
4) type of back-up system being used for temperature boosting.

The cheapest systems to install are obviously those requiring the least equipment. So combined collector/storage systems, like the 'breadbox collector' described in chapter four, must be high on the list for anyone doing the job on a shoestring. But the difficulty in properly insulating the storage tank in such a system means that they are really only appropriate for southern regions where the weather is warm and the sunshine uninterrupted.

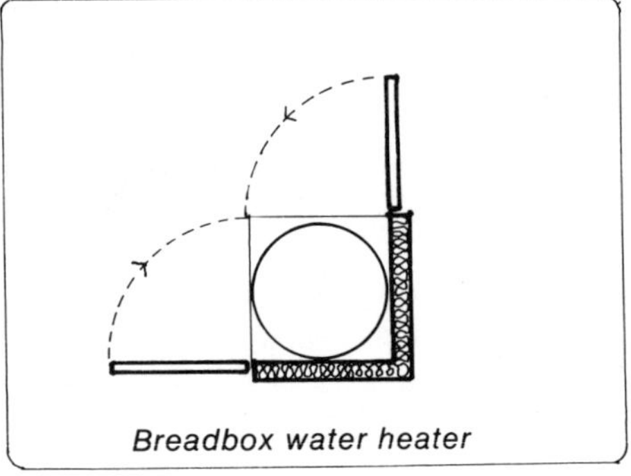

Breadbox water heater

If you decide to go for the more efficient set-up where the collector and storage tank are separated, a decision must be made about how to circulate water between them. As explained in chaper 5, there are two options: thermosyphoning and pumping. On cost-effective grounds thermosyphon systems are preferable, if suitable locations are available for mounting the collectors below the level of the tank without long pipe runs.

In areas where there is any danger of frost occurring, (and that means almost everywhere), it is essential to ensure that water in the exposed parts of the system does not freeze and thereby damage the

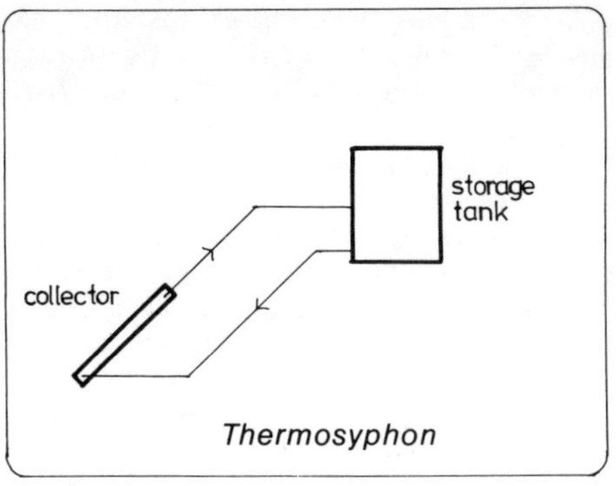

Thermosyphon

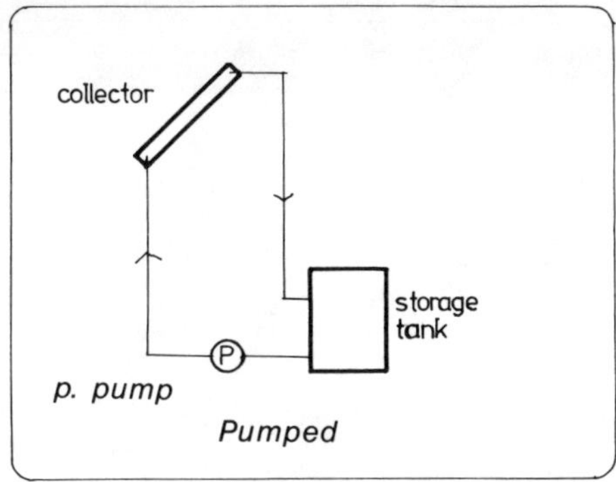

p. pump
Pumped

pipework. There are several ways of protecting the system.

Both thermosyphoning and pumped installations can be protected by adopting an *indirect system* of circulation. This is sometimes described as a *closed loop* system and necessitates the use of a heat exchanger to separate the water passing through the collectors from the service hot water which is delivered to the kitchen faucet thereby preventing its contamination. Once this is done, it is possible to use a non-freezing fluid instead of water to transfer heat from the collectors. If a closed loop is adopted, separate provisions must be made for venting air from the loop, and allowing for thermal expansion of the heat transfer fluid it carries. This usually means installing an automatic vent at the highest point, and a sealed expansion tank somewhere on the loop.

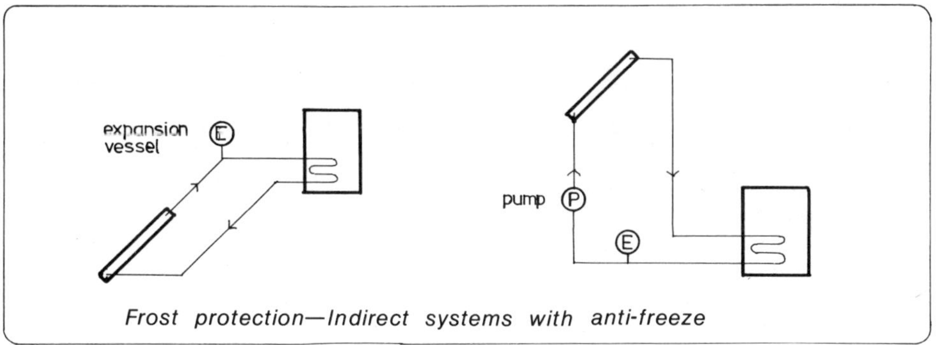

Frost protection—Indirect systems with anti-freeze

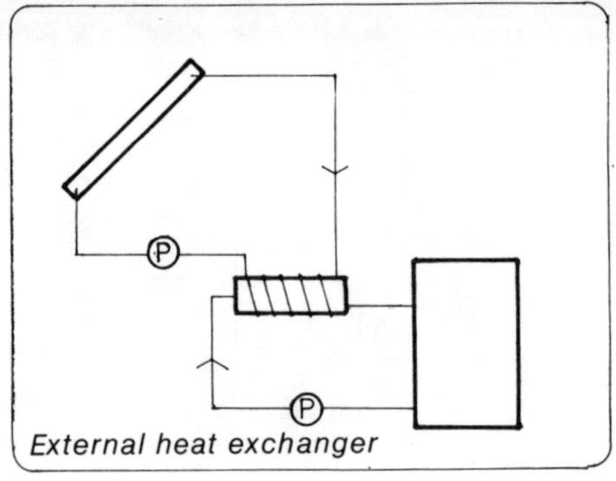
External heat exchanger

As explained in chapter six, heat exchangers can be integral with the storage tank, or separate from it. Separate heat exchangers require a second pump to circulate between the exchanger and the tank. This pump would be controlled by the same temperature differential device which controls the solar circuit pump. Separate heat exchangers are obviously inappropriate for thermosyphoning systems.

The use of any heat exchanger inevitably means a loss in efficiency. Only 60%–90% of the heat absorbed by the collectors will reach the water in the tank through the exchanger. A simpler and cheaper answer to the problem of freezing would be the use of insulating covers on cold nights. This however demands a good memory and self-discipline. To lay in bed on a sunny morning would more than ever be a waste of energy, whilst forgetting the covers on a cold night could be a costly mistake. Apart from these worries, moveable covers could only be a practicable proposition when the collectors are in an accessible position.

The only widely applicable alternative to the heat exchanger/anti-freeze solution is to drain the collectors whenever there is a danger of frost. Manual operation of valves is one obvious method of achieving this, but that raises all the fears of human foibles which we just considered as objections to the use of insulating covers. Automatic *draindown* can be achieved by installing solenoid valves which isolate the collectors from the tank and flush them whenever low temperatures threaten. This may mean

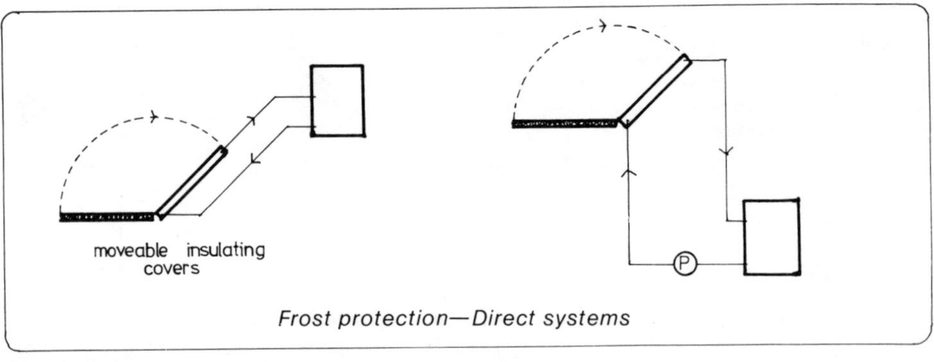

Frost protection—Direct systems

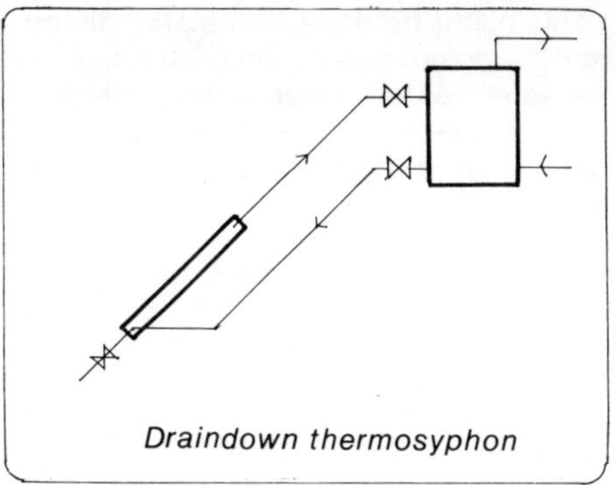
Draindown thermosyphon

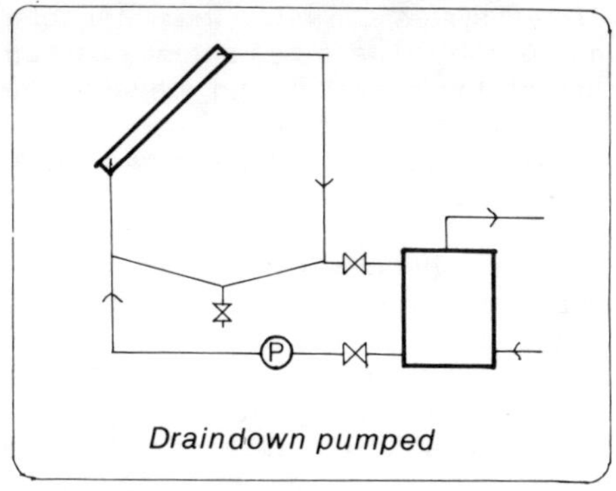
Draindown pumped

automatically draining every night in winter, or whenever a temperature sensor registers below a pre-set level. Such a draindown system could be fitted to a thermosyphon system but the necessary isolating solenoid valves would cause an undesirable addition to the flow resistance.

A *drainback* layout enables you to drain the system without wasting the water. Drainback is possible only in pumped systems. The collectors must be mounted above the storage tank so that the water flows back into the tank whenever the pump switches off. For this to happen, the water in the tank has to be open to air pressure and obviously cannot be connected directly to the mains supply. You must either decide to have a hot water system which is entirely low pressure, fed from a cold water storage tank mounted below the level of the collectors and filled via a ball valve, or you reintroduce the heat exchanger and have a low pressure solar circuit with a normal high pressure hot water distribution system to deliver water to the faucet. With the low pressure system, care must be taken to ensure that you create a big enough head of water to provide a strong flow at the outlet points. It would be annoying to install the system only to find that it could raise only a dribble of water from the shower head. As a rough guide, the level of water in the cold storage tank should be at least 5 feet above the highest faucet or shower head. With the high pressure system, supply water has to be heated at a single pass through the solar tank. This requires an extra large heat exchanger. The system suffers the efficiency loss associated with heat exchangers, but avoids the expense and trouble of using a non-freezing fluid.

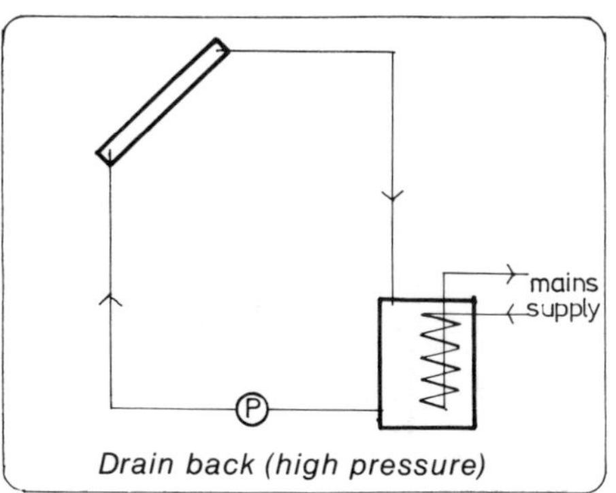

Drain back (high pressure)

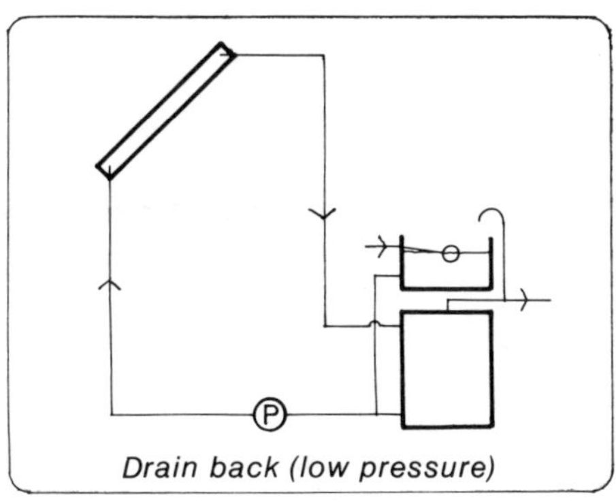
Drain back (low pressure)

Direct systems, in which the water coming out the hot taps has actually passed through the collectors and draining systems, are more liable to corrosion. Rusting for example occurs most readily where iron or steel comes in contact with water and air. There is always some air dissolved in water. In indirect systems, however, the amount of air present in the system is limited and once used up in initial surface rusting, it will not be replaced until the system is drained. In a direct system there is a continual flow of fresh water into the collectors and hence a continual supply of corrosive solution and mineral deposits. Direct systems should only be contemplated in installations where the collector is made of a highly corrosion-resistant material such as copper or stainless steel. In areas where there is a high concentration of carbonaceous compounds, as indicated by the presence of furring in kettles, an indirect system is advisable to avoid blockages in the pipe work.

Connection to Auxiliary Heating System

If there is an existing hot water system in the house, one could simply add a second hot water faucet for the solar heated water. A more sophisticated solution would be to integrate the two systems with the solar system acting as a pre-heater and the conventional system operating only when the solar heated water is not as hot as required.

This might be done with instantaneous heaters positioned at each outlet. When the solar heated water is hot enough to use, the heater could be turned off. When the temperature of slightly heated water needs boosting there are complications however.

Instantaneous gas heaters usually have controlling mechanisms which regulate the quantity of gas to be burned in relation to the volume of water flowing through the heater and not in relation to the temperature of the water. What is needed is a valve on the gas feed pipe controlled by a temperature sensor on the water feed pipe. It would be wise to seek the advice of the manufacturers before attempting such an alteration. Simple adjustment can be made by varying the rate at which you run the tap but this is a rather hit-and-miss methods.

Electrical instantaneous heaters often have a knob which allows one to vary the output temperature. This would be handy if it were to be fed pre-heated water. It is a simple matter to integrate the solar system with an instantaneous heater. The cold water supply is diverted from the unit and led into the bottom of the solar storage tank. The feed for the heating unit is then taken from the top of the tank. Warm water tends to rise. The top of the storage tank is therefore always the best position from which to draw off.

The attraction of instantaneous heaters is that they come into operation only when the hot tap is opened. There is no wastage therefore due to heated water being left

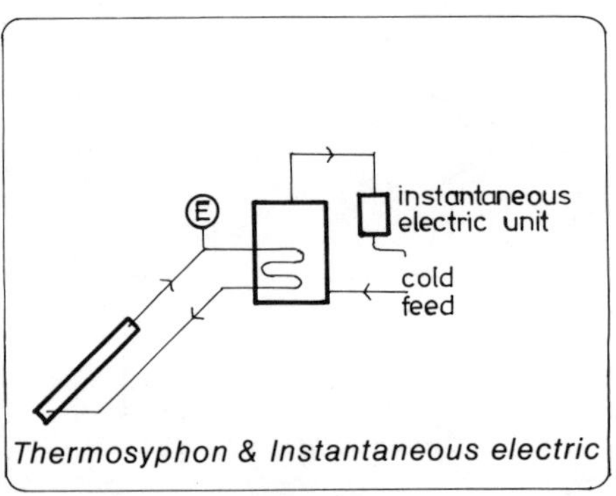

Thermosyphon & Instantaneous electric

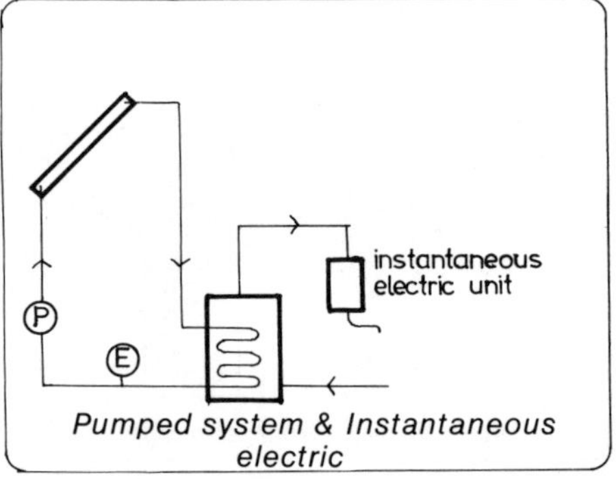

Pumped system & Instantaneous electric

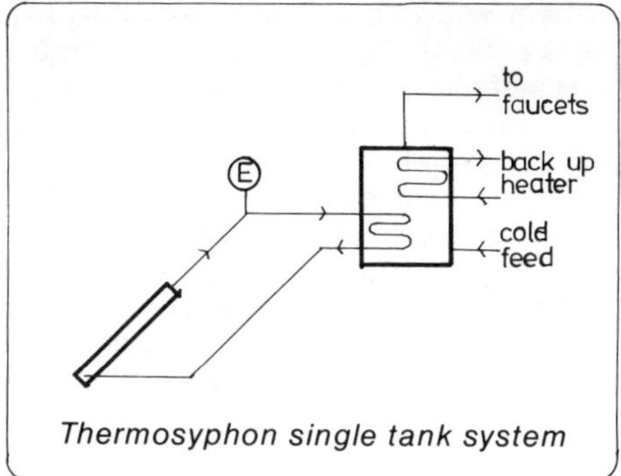

Thermosyphon single tank system

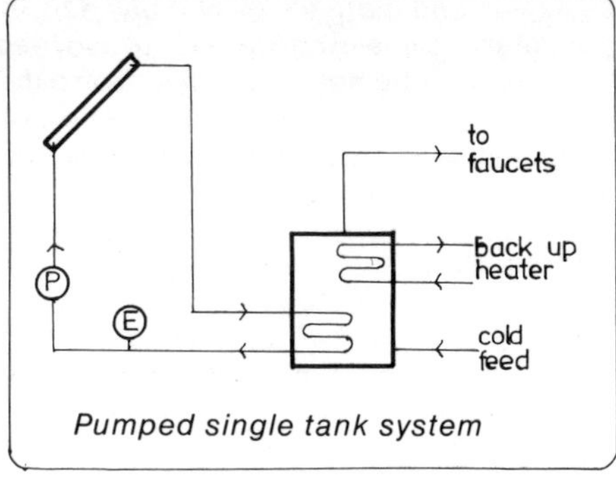

Pumped single tank system

unused. However, the difficulty in automatically adjusting their fuel consumption when operating with a pre-heated supply which varies in temperature means that storage heating systems ought to be considered.

The storage tank systems are heated by an electrical immersion element or by a separate boiler circulating hot water through the tank either directly or indirectly via a heat exchanger. The heat input to the tank is controlled by a manually operated switch or automatically by a thermostat or time switch. Solar preheating can take place in the existing hot tank or in a separate solar storage tank, which then feeds into the existing tank.

If the solar collectors are connected to the existing storage tank, there will obviously be savings in both spending and space. This system will work best with tall, vertical tanks in which a clearly defined thermal gradient will occur—hot at the top and gradually cooler towards the bottom. The conventional heating source must be placed in the upper half of the tank, above the solar input. This may mean the existing plumbing will need to be altered. Otherwise there is the danger that the whole tank will be heated in the normal way and the solar collectors will never become hotter than the tank temperatures. In such a case the solar system would never begin to circulate. If the conventional heater is raised to the upper half of the tank, this does of course mean that the hot water storage capacity, except in sunny periods, will be effectively halved. The single tank system will only perform satisfactorily then if the existing tank is a large size, e.g. greater than 60 gallons (270 litres) in an average size family house. Chapter 6 illustrates a means of connecting new heat exchangers to old tanks.

If one is dealing with an existing hot water system, then solar installation may

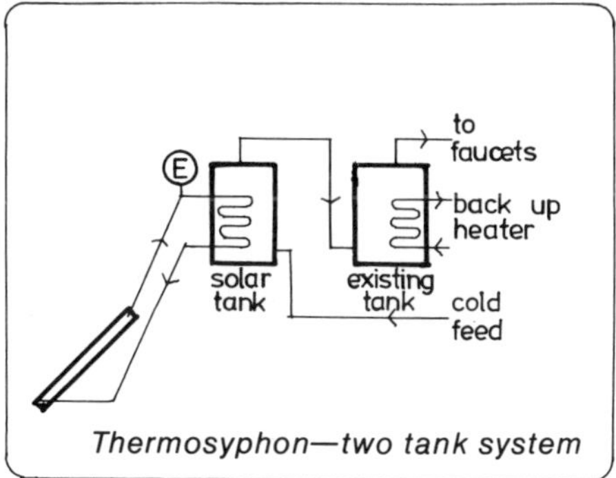

Thermosyphon—two tank system

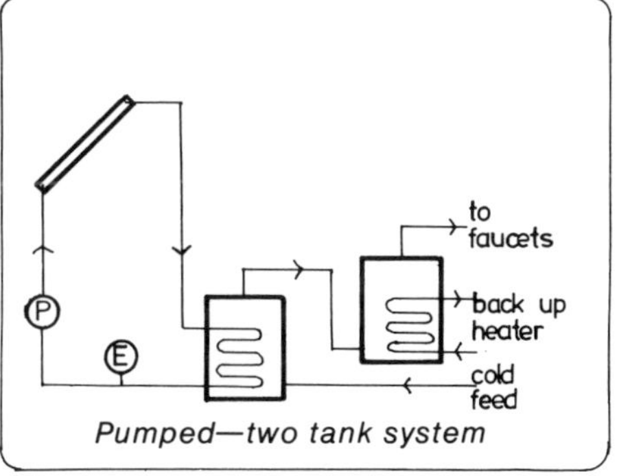

Pumped—two tank system

be simple and more efficient if one adopts a separate solar storage tank. The cold feed from the existing hot tank is diverted to the bottom of the solar tank which then feeds from the top, back to the bottom of the first tank.

Conclusion

This chapter has indicated the variety in system design. Other layouts which have not been shown may also be possible. The two tank system has been suggested as the most efficient for retro-fitting. It is important to study your existing water heating system before you decide on a solar system because obviously savings can be made by using what is already available to the best advantage. The next chapters describe plumbing in more detail and chapter 9 in particular might help you to understand better your existing heating system.

8

PLUMBING TECHNIQUES

The name "plumber" comes from the Latin word meaning lead, and that's what the original Roman plumbers used for everything. Piping, drainage, connections, everything was made on site from sheet lead.

Thinking of the task facing the ancient plumbers always makes me relieved to be living in an era where ingenious components bring simple plumbing work within the scope of just about anyone capable of wielding a wrench. Certainly the plumbing of a solar water heating system could be tackled by a home handyman or woman, given a knowledge of a few basic principles and techniques and an awareness of the components available.

This chapter describes the materials suitable for the job and illustrates the simplest, the conventional and the most economical techniques for connecting pipe to pipe, and pipe to tank. Readers with plumbing experience might proceed to the next chapters which indicate sizes and layouts, and examine the effect of different solar collector connection sequences.

Piping

For a long and trouble-free system life, you should use copper tube for all the pipework. There are two basic types of copper tube: rigid, drawn pipe, and soft annealed pipe. Rigid tube is available in 10 and 20 feet lengths (3 and 6 metres). It is generally used where pipework is to be exposed and a neat appearance is desired. Soft pipe comes in coils of various lengths which, because they are long, allow extensive runs of pipe to be installed without the trouble and expense of forming joints. It is also easier to work with if you are installing a system in an existing house and you need to snake the pipe in and out of tight spaces.

Rigid pipe cannot be bent. If you wish to make a bend with this material, you have to use a manufactured L-shaped fitting, known as an elbow, which is sweat soldered to the tube. If you have never sweat soldered, don't worry. It's not a difficult job, and there is a step-by-step guide to the process below.

With soft copper, large diameter bends can be made quite easily by flexing the pipe round a curved object. Plumbers often use their knee, but you may find it easier to cut a piece of wood to the required curvature or use a section of large diameter soil pipe to serve as a form. For small diameter bends a bending spring must be slid over the tube to spread the bending force and prevent the pipe from kinking. This is an important point because kinks in the pipe will increase resistance to flow and can lead to local corrosion. The problem with springs is that they are often difficult to remove after you have made the bend. The trick of the trade for loosening them is to bend the pipe slightly further than you actually require, and then pull it back to its intended shape. It also helps to twist the spring as you pull it off. Bending should always be a gradual movement as jerking movements will result in wrinkles.

There are three grades of pipe from which to choose. The thickest and most expensive is called Type K. It is usually used for underground pipework only. Type L is the most universally acceptable type but solar systems can also be constructed from the thinner and more economical Type M, if the local building code allows it. The higher cost of the thicker Type L might be justified if you are installing a draindown system or if your local water supply is particularly corrosive.

Jointing

There are five types of connections used in solar systems:

1. sweat soldered
2. flared
3. threaded
4. union
5. hose clamped

Sweat soldering is the most reliable and economic means of joining copper pipes. Flare fittings are easy to connect and can be useful when fitting a piece of equipment which may need future replacement as they can be disconnected at the twist of a wrench. Threaded joints are often used where a pipe connects with a tank. A union fitting is one with either sweat soldered or threaded connections, which can be split in two by loosening a nut in the middle. The nut can be turned independently of the fitting so that the fitting can be opened without disturbing the pipework on either side. For this reason it is also used when fixing equipment such as pumps which may require future replacement. Hose clamps are used to connect copper tube to short lengths of flexible pipe such as silicone coupling hose. Such hose couplings are used to prevent direct contact between dissimilar metals, to absorb thermal expansion in long runs of pipe, and in tight corners where there might not be space to turn a wrench or manage a soldering operation.

Making a Sweat Soldered Joint

A manufactured sweat fitting is a body formed as desired into a T-shape, an L-shape or a straight coupling. Each end has an opening like a sleeve into which the pipe fits tightly. To secure and seal the joint, solder has to be run into the narrow crack between the contact surfaces. This is not as difficult as it might seem because capilliary forces tend to draw molten solder into the joint if it is fed to the surface of the crack.

To do this you will need a propane gas torch, some steel wool or cloth-backed emery paper, a hacksaw or tubing cutter, a half-round file, a spool of solid core solder wire (50:50 tin/lead), and a tin of compatible flux paste (preferably water-based). If you are doing the job for the first time, it is worthwhile making some experimental joints then sawing them open to see how successful you have been in sealing the joint with solder. Copper is a very good heat conductor, (remember chapter 3 and 4), so the whole tube may get very hot while you are making the joint. Let it cool for several minutes before you touch it. Holding the work in a clamp is a good way of avoiding blisters on your fingers, and you might keep a bowl of cold water handy — just in case. Here we go:

1. Cut the pipe to the desired length with a hacksaw or pipecutter. The cut should be made at right angles to the run of the pipe.
2. If the pipe end has been bent inwards while cutting, open it out again with a half-round file. Care should be taken not to flare out the end of the pipe such that it might not fit in the sleeve of the fitting.
3. File off any burr from the end of the pipe.

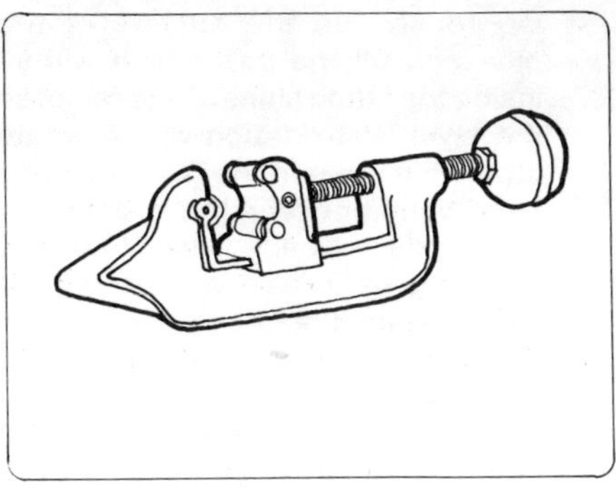

8.1 Tubing cutter

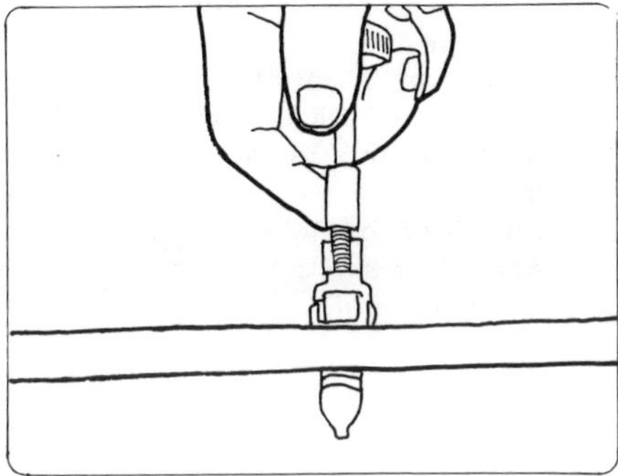

8.2 Using a tubing cutter: revolve round pipe, while slowly increasing blade pressure by twisting handle knob

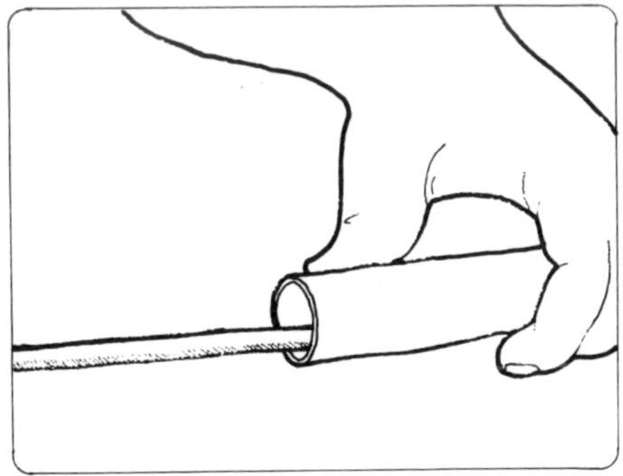

8.3 Removing burr with round profile file

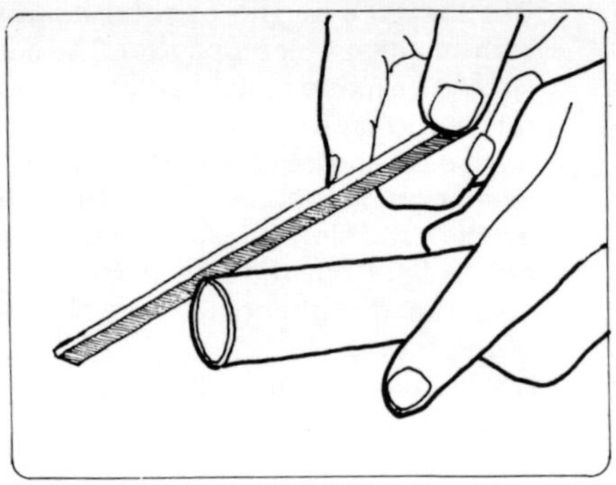

8.4 Removing rough edges with file

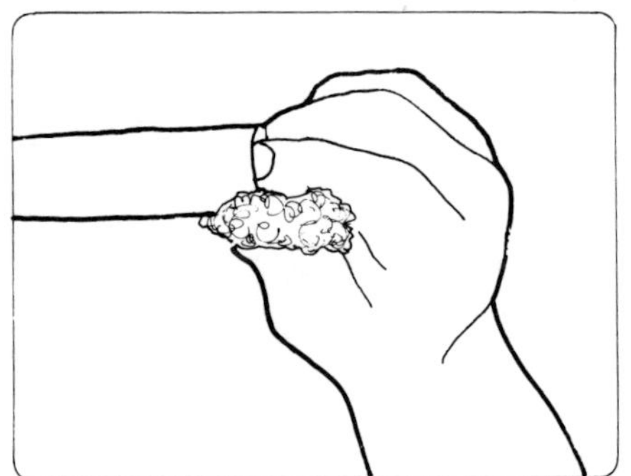

8.5 Brushing pipe end clean with steel wool

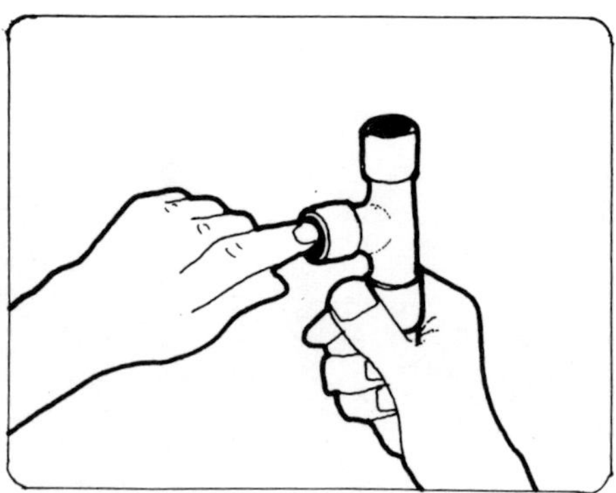

8.6 Smearing flux on the contact surfaces after cleaning with steel wool

4. Vigorously rub the *outside* of the pipe end till the part which will sit inside the fitting shines. This removes the layer of oxidation which would otherwise prevent the formation of a good bond between the copper and the solder. Check also that the copper is not greasy and use cleaning solvents if necesary.
5. Similarly brush and clean the *inside* of the fitting.
6. Smear the pipe and the inside of the fitting with flux. A small brush will be useful here. The idea of the flux is to help the solder flow smoothly and to prevent the cleaned copper from re-oxidising when it is heated with the torch.
7. Insert the pipe into the fitting and push it in till it comes to rest on a shoulder about half an inch (15mm) inside.
8. All other branches on the fitting should be similarly prepared before applying heat.

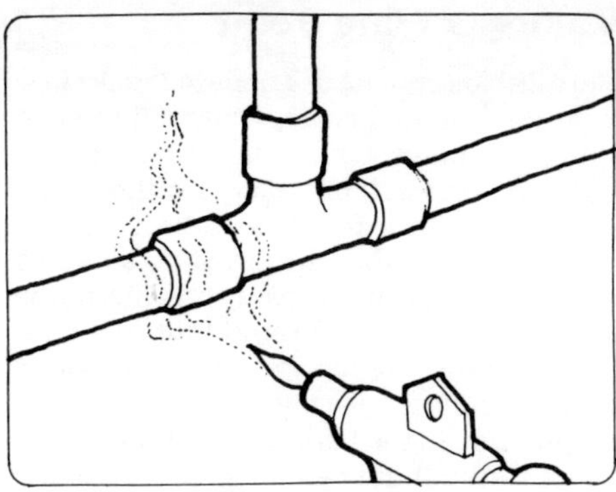

8.7 Heating the assembled joint with a propane torch

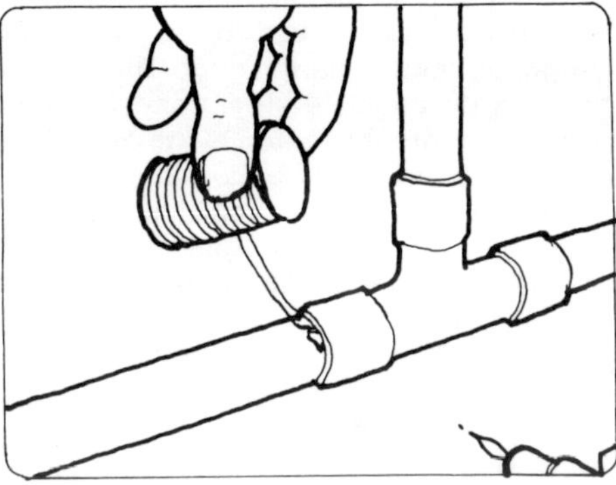

8.8 Applying solder to the heated joint

9. Using the propane torch apply heat; first to the tube, then alternate between the tube and the fitting.
10. Take the solder wire in your free hand, and dip its end in the tin of flux.
11. After a minute or so the colour of the copper will begin to change. This is an indicator that it may be sufficiently hot to begin applying solder. Test its readiness by touching the end of the solder wire against the copper. If it melts on contact, the copper is hot enough to begin feeding solder into the joint.
12. Alternate between heating the joint with the torch, and feeding solder into the mouth of the fitting. Never hold the solder wire in the flame. If it does not melt on contact with the copper, reheat the copper not the solder wire. Whist reheating the copper, dip the wire into the flux. This will help capilliary attraction draw solder into the joint.
13. When a bead of solder is formed round the whole circumference of the joint, the job is complete.

Notes:

a. After the tube and fitting have been cleaned and fluxed, the soldering should be done within about one hour.

b. The joint can be disassembled simply by reheating until the solder is molten and then pulling the pipes apart.

c. Once disassembled a fitting can be reused after cleaning.

d. If it is necessary to solder one branch of a fitting before the other branches are prepared, care should be taken to prevent disturbing the first joint when forming the second. A wet rag must then be wrapped around it in order to absorb the heat and prevent the solder in that branch from melting.

e. Bending a pipe after it is soldered can weaken the joint. Any bending should therefore be done before soldering.

Making a Flare Joint

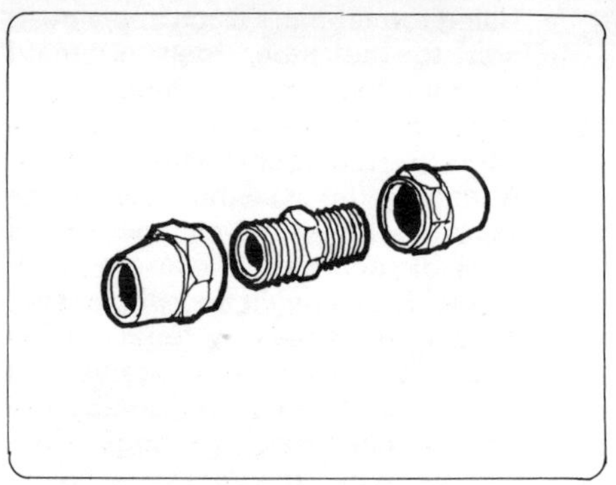

8.9 *Flare joint fittings*

Flare fittings consist of a body in the desired form, (L-shaped elbows, T-shaped tees, or straight couples) usually made from brass, with nuts on each outlet where the fitting connects to a pipe. The outlets are so machined that they fit snugly inside the mouth of a flared copper tube. The nut is there to hold the tube and fitting in tight contact. When tightened up, these joints are self sealing. They can only be used with soft copper as the flaring operation cannot be carried out successfully on rigid pipe.

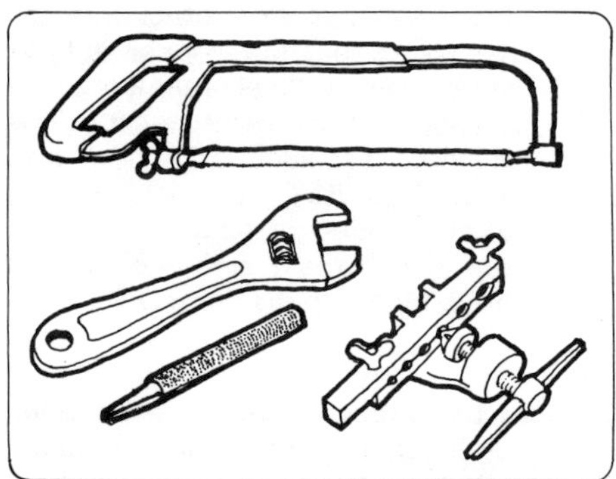

8.10 Tools for making joints: hacksaw, file, wrench and flaring tool.

To make the joints you will need the following: a pipecutter or a hacksaw, a reamer and/or a half-round file, two wrenches, and a flaring tool. The flaring tool can be a simple steel cone which is hammered into the mouth of the pipe, or in a more sophisticated version, it is a clamping device in which the tube is held while a cone on the end of a threaded rod is screwed into it. Let's go through one step at a time:

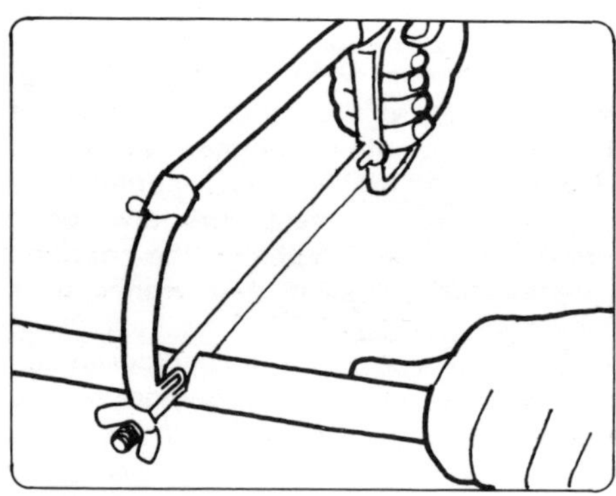

8.11 *Cut pipe to desired length*

1. Cut the pipe to the desired length using a hacksaw or pipecutter. The cut should be at right angles to the run of the pipe.

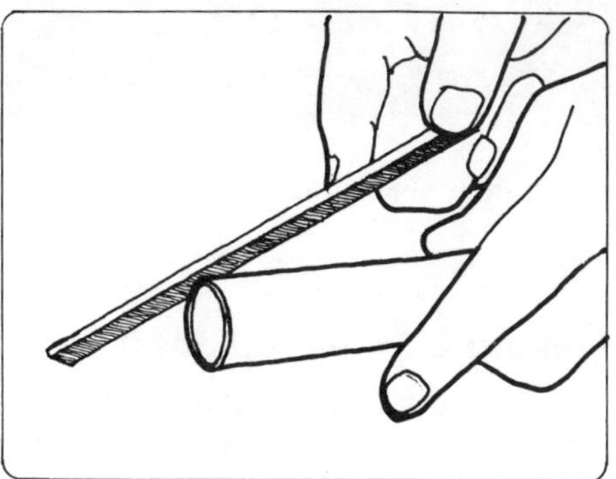

8.12 *File off burrs*

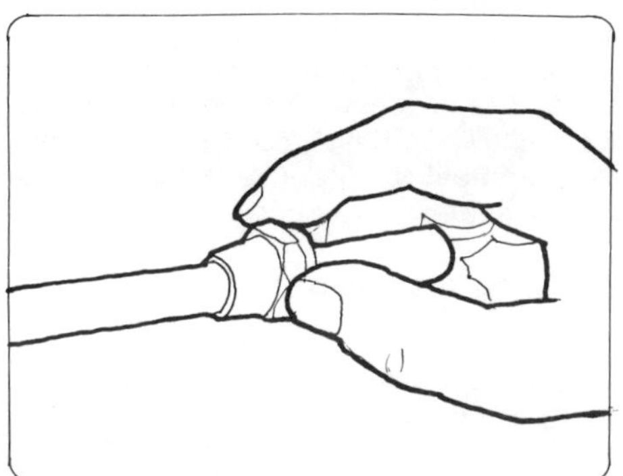

8.13 *Slide nut onto pipe*

2. File off, and ream out any burr left on the pipe end.
3. Loosen the nut on the fitting by turning counter-clockwise, and remove it.
4. Slide the nut over the end of the pipe. You must do this *before* flaring the pipe.
5. Flare out the end of the tube by clamping it in a flaring tool die, so that the end projects slightly above the face of the die. Then tighten the handle on the flaring tool so that the cone enters and flares the tube.
   or
   Flare out the tube end by carefully hammering in a flaring tool.
6. Inspect the flared end to ensure that it is circular, even and uncracked. The flaring tool can be re-applied if the tube is uneven but if it is cracked, saw off the end of the pipe and begin the flaring operation from scratch.
7. Push the body of the fitting into the flared tube end, slide up the nut on the tube, and tighten using two wrenches.

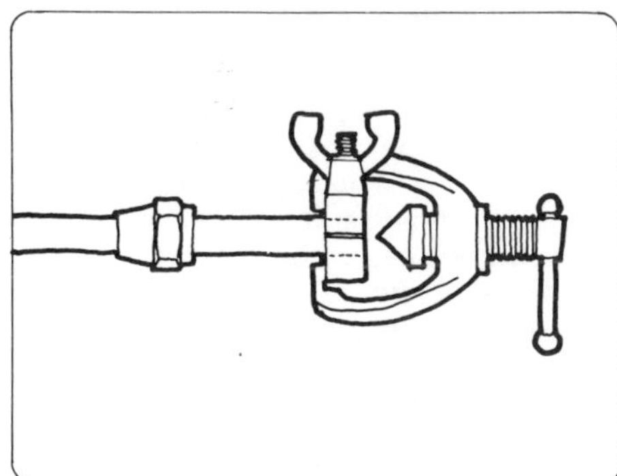

8.14 *Flare out end of tube using flaring tool*

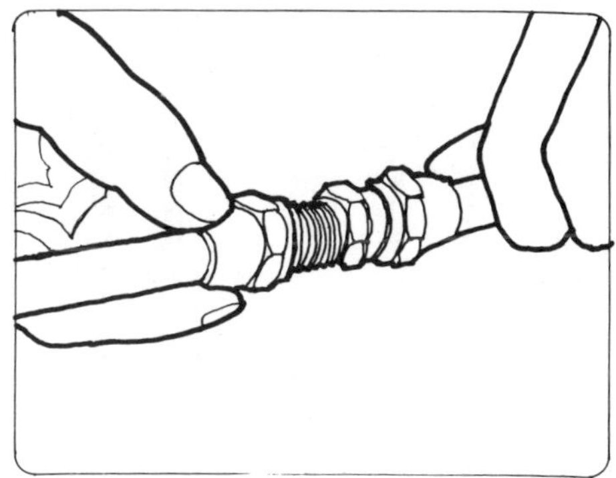

8.15 *Push body of fitting into flared tube & secure with nut*

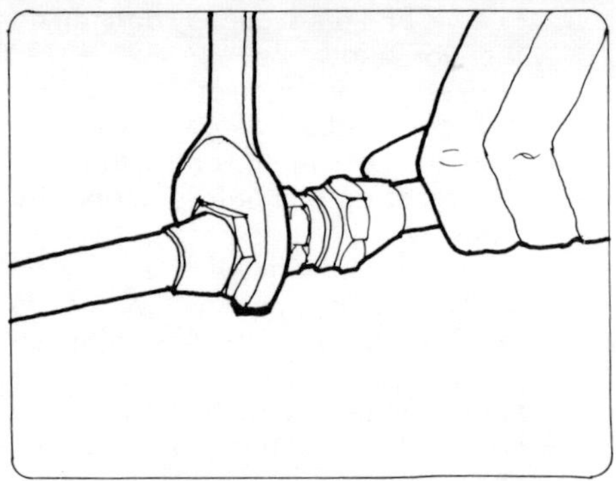

8.16 Complete flared joint by tightening nuts with wrench

8.17 Teflon® tape is used to seal thread joint

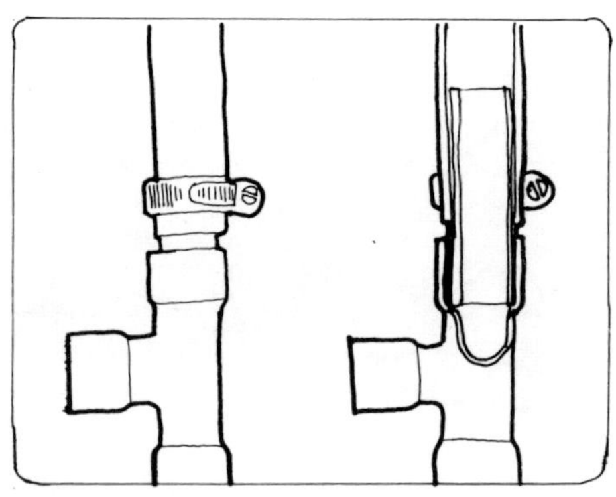

8.18 External and cross section view of copper/high temperature rubber joint.

Making a Threaded Joint

These are used mostly where pipes connect with a tank. The threads are distinguished as being either external (male) or internal (female) threads. The watertightness of the joint depends on the prevention of water seeping through its threads. This is achieved by the use of a sealing paste and/or Teflon® tape which is wound round the external thread. Always use a good quality pipe compound such as Loctite® pipe dope with Teflon (Part No. 529), Dow Corning® 730 Fluorosilicone, Rector® Seal No. 5 or Crane ® JE 30 pipe thread sealer with Teflon®.

Making a Union Joint

A union fitting comes in two pieces, one of which has a nut attached which holds the two parts together when tightened. This nut-tightened joint is self-sealing, but the connection of the fitting to the pipe is usually either sweated or threaded, and the procedures outlined above should be followed when fixing a union fitting.

Making a Hose Clamp Joint

This joint should be used only for joining copper tube to high temperature rubber tubing. The tube must also be ultraviolet resistent if it is to be exposed to sunshine. Automobile hose is not good enough.

The rubber has to be pushed over the end of the copper tube forming a tight fit. This joint is fixed and sealed by tightening the screw on the hose clamp. Use stainless steel clamps without perforations.

Plumbing Principles

There are three principles involved in plumbing solar systems which if properly appreciated will help you work out your plumbing requirements. These are: Water finds its own level; Air is dissolved in water and it comes out of solution when heated and floats up to the highest position it can reach; and warm water tends to rise above cold water in convection currents.

Water Levels

If you connect a pipe to a tank full of water, and take that pipe round an obstacle course of bends, vessels and other tanks, the water flowing into the pipe will always try to reach the same level as the surface of water in that first tank. If the end of the pipe is held below this level, there will always be pressure there, unless the tank is allowed to empty itself through the pipe and thereby equalise its level on the ground. Only when the pipe end is lifted above the tank water level will the pressure cease and hence water would stop flowing out through it. This is important to remember if you are designing a drainback system. You won't want to have water trapped in the bottom of your collectors.

Dissolved Air

To continue with the example of a pipe connected to a tank of water, if the pipe end was held below the level of water in the tank and water did not flow out through it and there were no kinks in the pipe, we might be almost certain that there were air blockages in the pipe. Air is present in pipework before it is filled and can also appear after filling if the water in the pipe is heated. This happens in the same way that air bubbles appear in a saucepan of water being boiled. Because the air bubbles are less dense than water, they attempt to float up to the water surface. In plumbing systems, however, they are often trapped by bends in the pipework. Such points where air is trapped must be fitted with air vents, either automatic or manual, so that the air does not form a blockage to the water circulation.

Convection

The tendency of warm water to rise above cold water has been mentioned several times and is relevant to the working of gravity circulation systems and the locating of hot water drawoffs.

Conclusion

This chapter has described the techniques and principles you need to know to tackle your own plumbing. The next chapter will detail the procedures involved in putting your new knowledge into practice in the installation of a solar water heating system.

9

PLUMBING A SOLAR SYSTEM

It would be fine and simple if we could just say–run pipes from the solar collectors to the storage tank, and then the faucet. That is exactly what we have to do, but in the process we must also consider what will be happening inside the pipework and the practical requirements of the installer and the householder.

Inside the pipework we have to think about:

—the expansion of heated water,

—the removal of the air which bubbles out of heated water.

Outside the pipework we have to plan for people who need

—safety precautions to prevent temperatures and pressures becoming dangerously high, and to maintain the purity of their water supply,

—simple methods of filling and emptying the pipes,

—simple maintenance procedures.

These factors may involve us in adding lots of valves and branches of pipework but if the diagrams remind you of spaghetti, don't despair, remember that all we are actually doing is to connect the collectors to the tank, and the tank to the faucet.

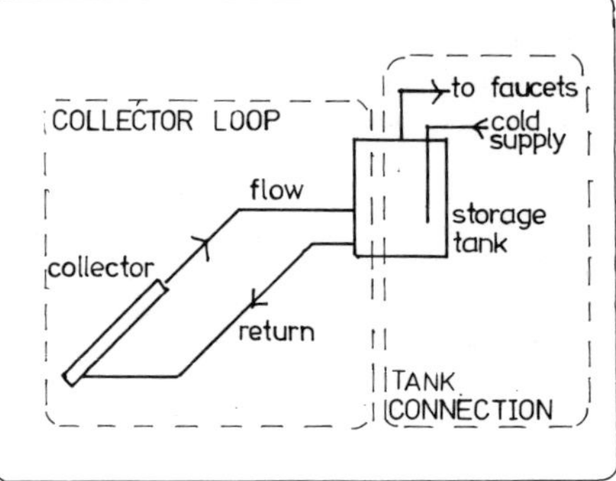

9.1 Collector loop/Tank connection

In this chapter we will look in detail at how three different types of system are plumbed. We start with a simple thermosyphon system, then look at a pumped closed loop and finally, examine the layout of a draindown system. Having read chapter 7 you may have already decided on which system you want, but nonetheless it will probably help to have a look at the details for all three types.

9.1 Thermosyphon—open loop system
Key
a. Shutoff (isolation) valve
b. Strainer
c. Pressure reducing valve
d. Temperature & pressure relief valve
e. Boiler drain valve
f. Dip tube

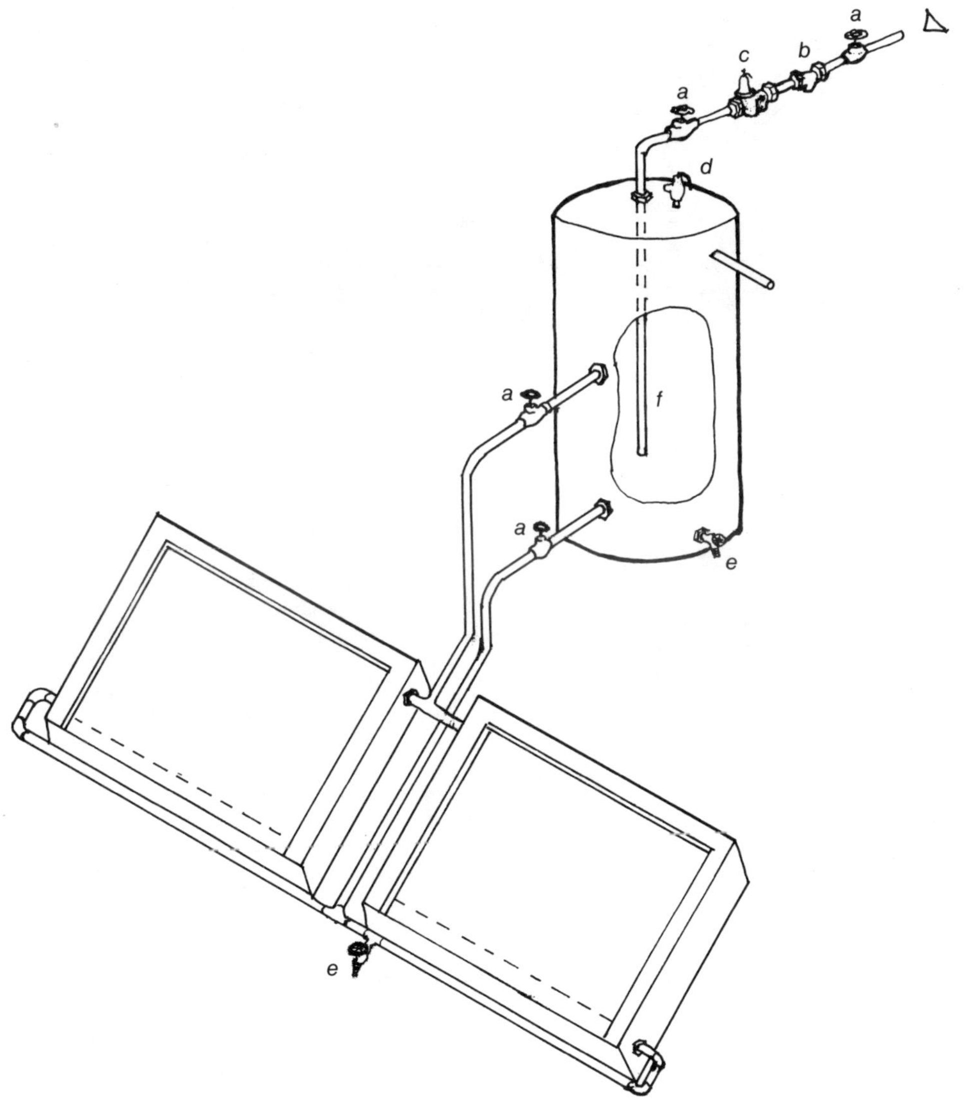

9.3 *Boiler drain valve (e)*

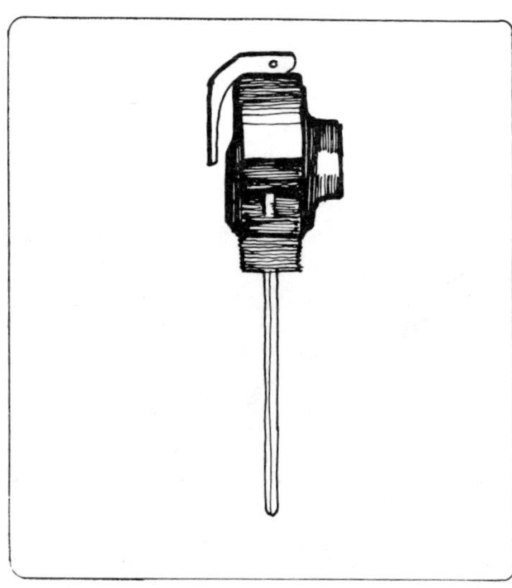

9.4 *Temperature & pressure relief valve (d)*

Thermosyphon System

In a thermosyphon system the overriding requirement is that we minimize the resistance to water flow between the collectors and the storage tank. In practical terms this means using large diameter pipes (preferably 1" or bigger) and minimizing the length of piping by locating the collector and tank as close together as possible. We should also have a continuous rise in the pipes from the collector up to the tank, and avoid any unnecessary bends.

The more fittings and valves that are added to the circuit, the more resistance there will be to water flow. In open loop thermosyphon systems all that is required is a drain valve at the lowest point in the circuit, and two shutoff valves to make possible the draining of the collectors without emptying the whole tank. (That can be an essential feature in climates where draindown is necessary in winter to prevent freezing.) Note that with this layout the collector loop is subjected to water supply line pressures. Care should be taken therefore to ensure that any joints and components used can withstand pressures of up to 100 p.s.i. Alternatively a pressure reducing valve should be added to the supply line at some point before it enters the hot water tank. The valve should have a strainer installed upstream of it, and shutoff valves on either side to facilitate cleaning. The storage cylinder should have a drain valve fitted at the bottom and perhaps most important of all, at the top it must be fitted with a temperature and pressure relief valve which is usually rated at 210°F and 125 p.s.i.. This valve must be connected by a copper tube to discharge over a drain. In an open loop system this valve will prevent pressures in the collector pipework growing dangerously high.

9.5 Thermosyphon—closed loop system
Key
a. Shutoff (isolation) valve
d. Temperature & pressure relief valve
e. Boiler drain valve
f. Dip tube
g. Automatic air vent
h. Pressure relief valve
i. Expansion valve

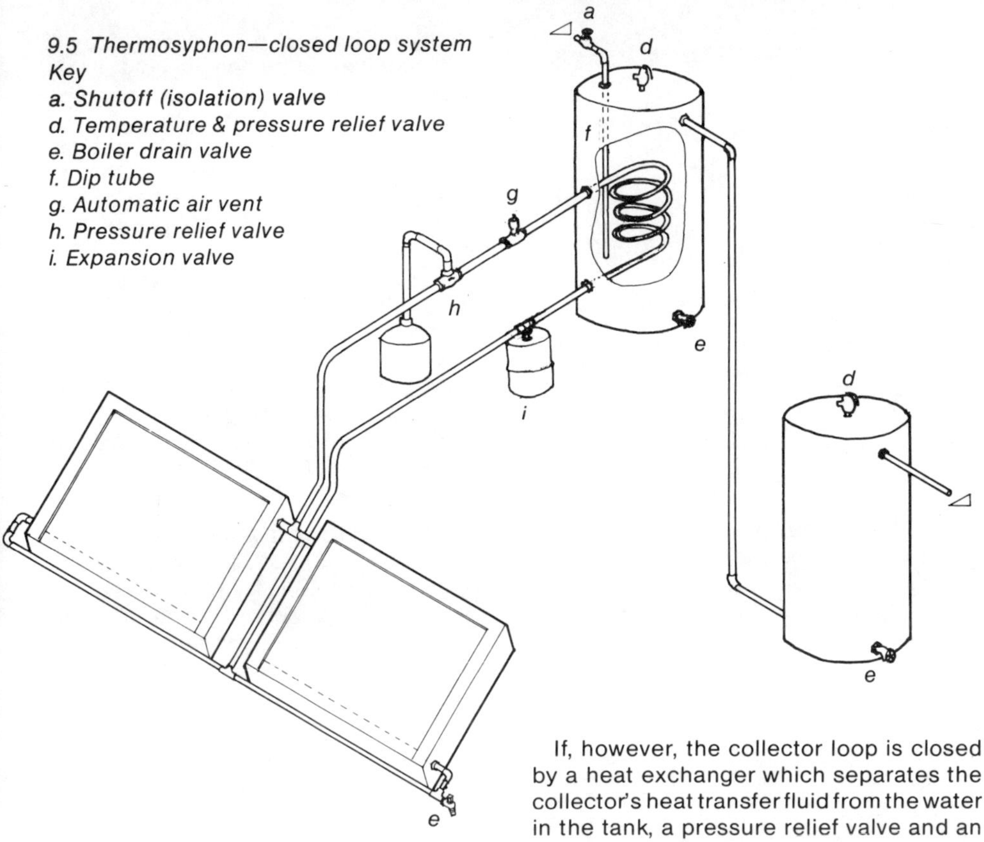

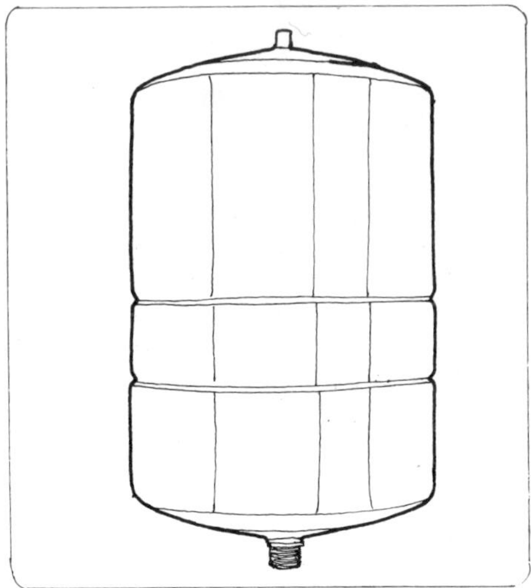

9.6 Expansion tank (i)

If, however, the collector loop is closed by a heat exchanger which separates the collector's heat transfer fluid from the water in the tank, a pressure relief valve and an expansion vessel will have to be added to the pipework. A sealed expansion tank can be placed anywhere on the circuit, but make sure that the relief valve is positioned where there are no other valves between it and the solar collectors. This is to ensure that water in the collectors, which might begin to boil, will never be isolated from this safety valve. An air vent also becomes necessary to prevent air blockages in the pipework. The automatic float type vent is the best choice, and it should be mounted on a tee fitting at the highest point in the circuit; i.e. where the hot 'flow' pipe, coming from the top of the collector, enters the heat exchanger. It is important that the highest point should not be inside the exchanger, as might happen if the tank were mounted on an uneven base with the exchanger inlet on the lower side of a tilted tank.

9.7 Pumped closed loop system
Key
a. shutoff valve
b. Strainer
d. Temperature & pressure relief valve
e. Boiler drain valve
f. Dip tube
g. Automatic air vent
h. Pressure relief valve
i. Expansion tank (sealed)
j. Air eliminator
k. Pump
l. Globe valve
m. Pressure gauge
n. Check valve
o. Backflow preventer
p. Mixing valve

Unlike the open loop, the closed loop will not fill automatically when the cold supply to the cylinder is turned on. It has to be filled under pressure via the drain valve at the bottom of the system, or by pouring liquid in from the highest point; i.e. the tee on which the vent is mounted. For details see the paragraph on 'flushing and filling' below.

With both varieties of thermosyphon systems there is a possibility that heat which has been collected during the day may be lost at night due to reverse circulation from the warm tank to the cold collector. This can be prevented by ensuring that the bottom of the tank is mounted at least two feet above the top of the collector.

Pumped Closed Loop System

The increase in sophistication, and expense incurred, when you opt for a pumped system is expressed clearly enough in the increase in the number of gadgets which can be seen bristling from the pipework in figure 9.7.

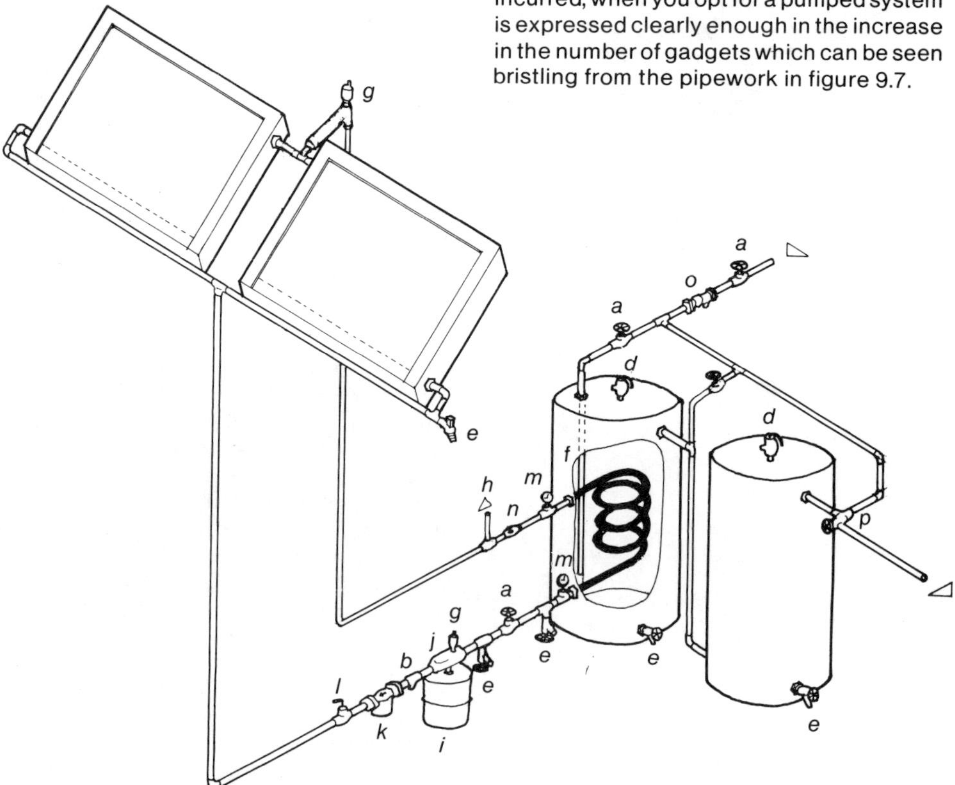

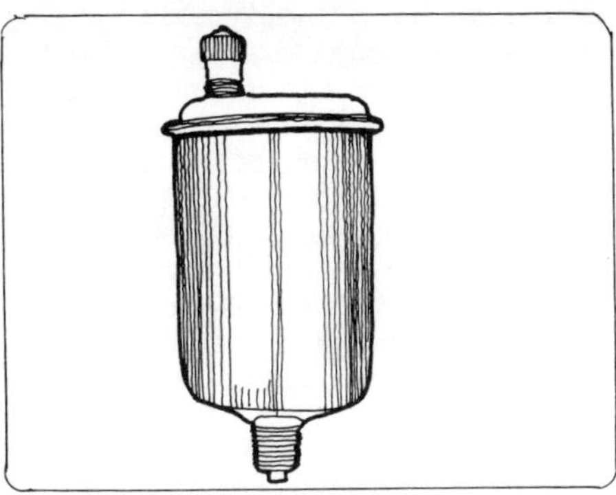

9.8 Automatic air vent (g)

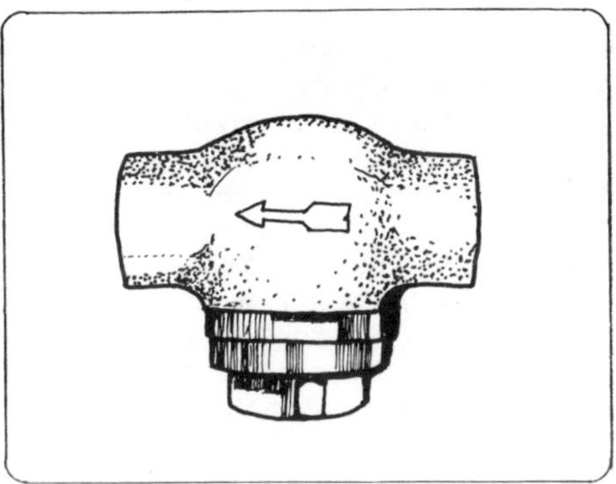

9.9 Check (one way) valve (n)

9.10 Pressure relief valve (h)

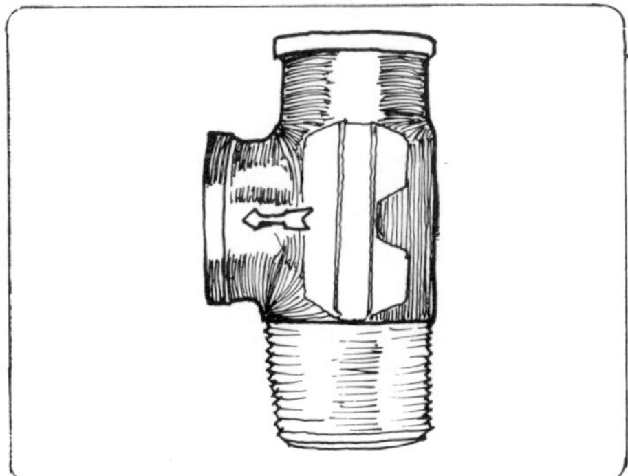

Smaller diameter, and therefore cheaper, pipe can be used in pumped systems though. Usually ½" is adequate, but ¾" would be used for extra large arrays of collectors, or when the pipe runs stretched out beyond something like 100 feet.

Let's start at the top of this system and work our way round. At the highest point, as you have probably guessed by now, we will find an automatic air vent. In pumped systems it is worth taking a little extra care when installing the vent. Air will escape most readily if the velocity of the water circulating immediately below the vent is slowed down. This can be achieved by changing to a larger diameter of pipe a short distance upstream of the vent.

Following the 'flow' pipe from the collector down to the cylinder, we come first to a pressure relief valve. This will discharge steam if the collector becomes so hot that normal working pressures are exceeded. Standard valves are set to blow at between 30 and 45 p.s.i. Its discharge should be piped to a drain or container and should be open to the atmosphere. There should be no shut-off valves between this safety valve and the collector. In large systems where there are several banks of collectors separated by valves, each bank of collectors must have its own relief valve.

Moving closer to the cylinder, we come to a check valve, also known as a one-way valve. This is required to prevent reverse circulation which can occur at night due to the natural tendency of water to rise from a warm tank to the collector where it is cooled and then syphoned back to the tank. Reverse circulation won't do much to heat the night sky, but it will certainly cool the contents of your hot tank and needs to be avoided. The direction of flow is indicated by an arrow on the body of the valve. Make sure this is pointing in the direction of the tank, otherwise night-time will be the only time the system circulates. If possible the check valve should be mounted on a horizontal length of pipe and should be the 'swing' type. If the valve has to be mounted on a vertical pipe, use the spring type instead.

Somewhere along the line we need a

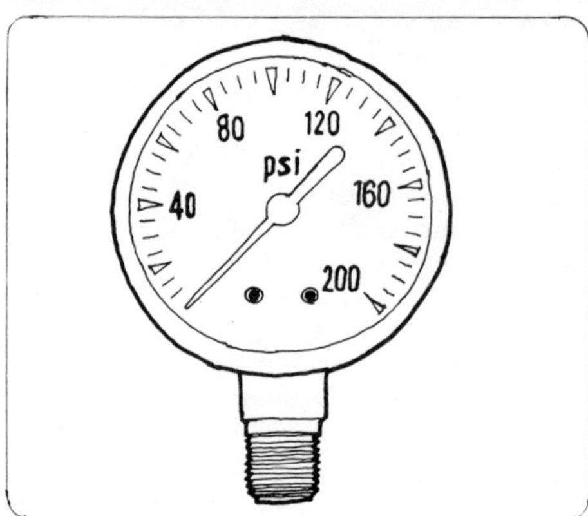

9.11 *Pressure Gauge (m)*

pressure gauge. It should be somewhere it will be seen as it is the main indicator for giving warning if the systems develops a leak, or if the diaphragm in the expansion tank has punctured, or the fluid has boiled and discharged through the relief valve. All these events would be signalled by a loss of pressure.

Sometimes the pressure gauge is combined with a temperature gauge. It is not essential to the system's operation that you know how hot it is, but it may well be of interest to you. It can be of use when you are adjusting the flow rate. The idea here is that you control the flow rate to give a temperature rise of between say 5 and 12 degrees F, in the water coming out of the collector. To measure you obviously need two sensors, and for convenience these are often located at the inlet and outlet of the heat exchanger.

Close to the tank is usually also a suitable and accessible position for the filling assembly which consists of two drain valves separated by a gate valve. The use of these valves is explained below under the heading 'flushing and filling the system'.

If the circuit is pressurized sufficiently the sealed expansion vessel can be placed anywhere on the collector loop. The pressure should be at least 5 p.s.i. at the top of the system. If the gauge is at the bottom of the pipework however it will also be registering the static head created by a column of water and the required pressure should therefore be higher; e.g. 15 p.s.i. at the bottom of a system with an overall height of 20 feet. The surest place to install the expansion vessel is on the suction side of the pump. It is normal practice to suspend it on a tee fitting below a horizontal length of pipe, in combination with an air separator and an air vent. The separator is a baffle which slows and churns the current of water encouraging dissolved air to bubble out.

The pump itself is best preceded by a strainer which will prevent dirt, lumps of solder and other debris from entering and damaging the most valuable component on the circuit.

When installing the pump, follow closely the makers' instructions. These should specify whether it can be mounted on vertical and/or horizontal pipes and which way up its motor head should be set. There will be an arrow on the body of the pump to help ensure that you get the water to circulate in the direction you want it to go.

Downstream of the pump, a globe valve serves the double function of allowing the installer to adjust the flow rate (assuming there are no such controls on the pump itself), and, in conjunction with the gate valve in the filling assembly, it enables you to isolate and service the pump without draining down the whole circuit.

9.12 *Air eliminator (j)*

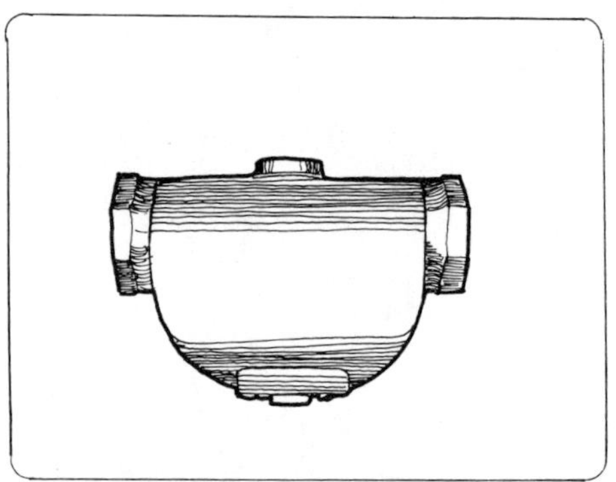

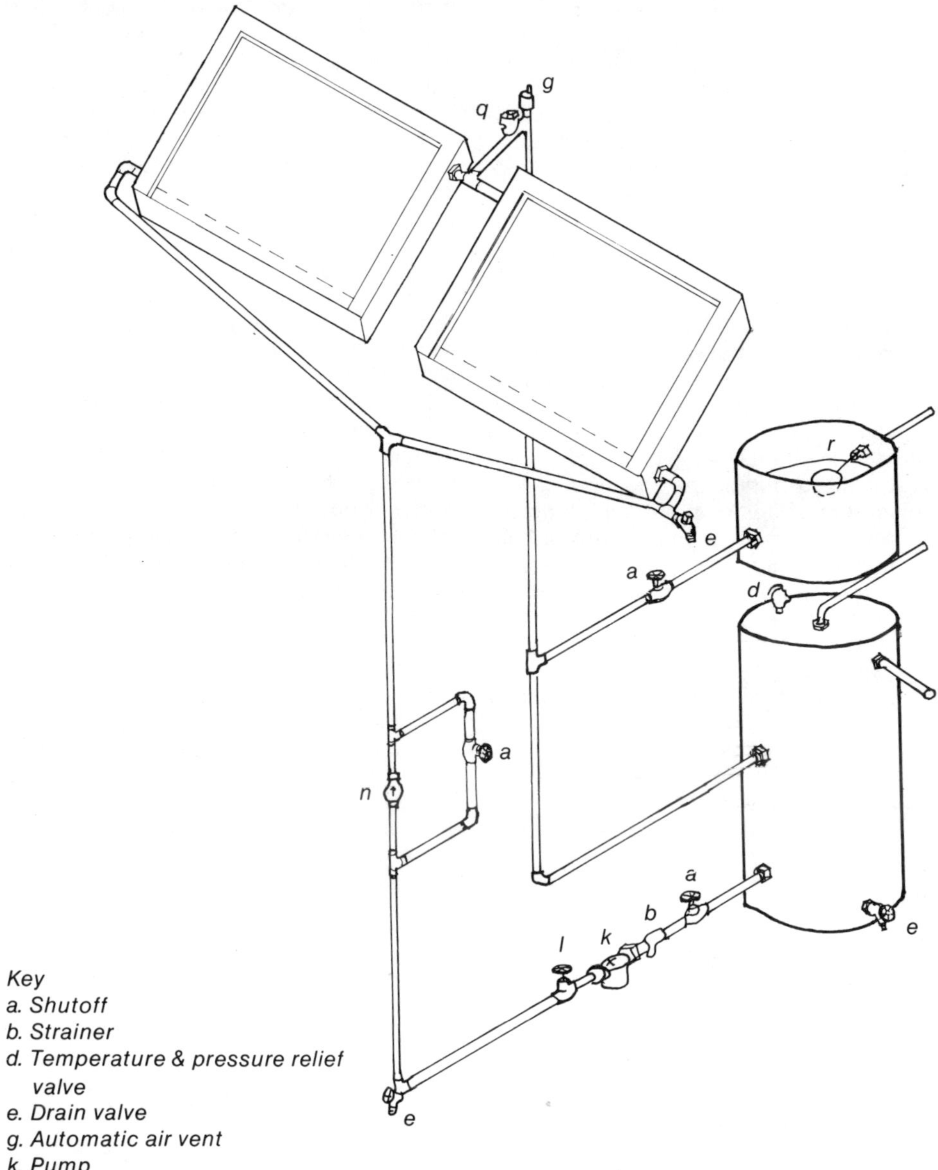

Key
a. Shutoff
b. Strainer
d. Temperature & pressure relief valve
e. Drain valve
g. Automatic air vent
k. Pump
l. Globe valve
n. Check valve
q. Vacuum relief valve
r. Float valve

9.13 Pumped Drainback System

Drainback System

The drainback system protects itself from frost and boiling by emptying the collectors whenever such dangers arise. As in the thermosyphon system, special care must be taken to ensure that all pipes slope steadily downwards from the highest point. There must be no dips in the pipework which could prevent drainage.

At the top of the system, in addition to the familiar air vent, a vacuum relief valve is also required to admit atmospheric pressure when required so that the system can drain under the influence of gravity. An alternative to the float vent and the vacuum relief is an open vent. This could simply be a length of ½" pipe projecting vertically above the collectors and bent over at the top in an inverted U-shape. The open vent has the disadvantage that the more care is required in balancing the pumping power to prevent over pumping and loss of fluid through the open pipe. The open vent will also increase evaporation losses. The subsequent need for frequent topping up of the water level has the attendant problem that the concentration of impurities in the water will slowly build up. For this reason it is recommended that they be charged initially with distilled water.

No additional valves are required on the pipe which runs from the top of the collector down to the cylinder. There is however a branch off to the open top tank which receives the water draining out of the collectors, and a gate valve should be fitted to this branch line. This tank should have a tight-fitting lid to reduce evaporation losses and it should be connected to the water supply line via a float valve which will automatically replenish any fluid loss which does occur. The water level in this tank determines the level to which the pipework drains. It must therefore be below the level of any exposed piping. But, the further below the collector it sits, the greater will be the pumping power required to refill the collectors. This tank also serves as the filling point.

Somewhere near the outlet of the heat exchanger, will be the lowest point in the system, and a drain valve should be fitted there to facilitate maintenance. This should be followed by a gate valve and another drain valve to form an assembly which can be used to flush the system as described above. (When flushing this system, remember to close the circuit by closing the gate valve which isolates the open top tank.)

Next comes a strainer, the pump and a globe valve. Pumps in draindown systems must be stainless steel or bronze. In fact it is in many ways desirable to install two pumps in series on a drainback system. It requires much more power to refill the system than to keep it circulating once

9.14 Drilling hole in tank, fitting float valve in hole and tightening back nut

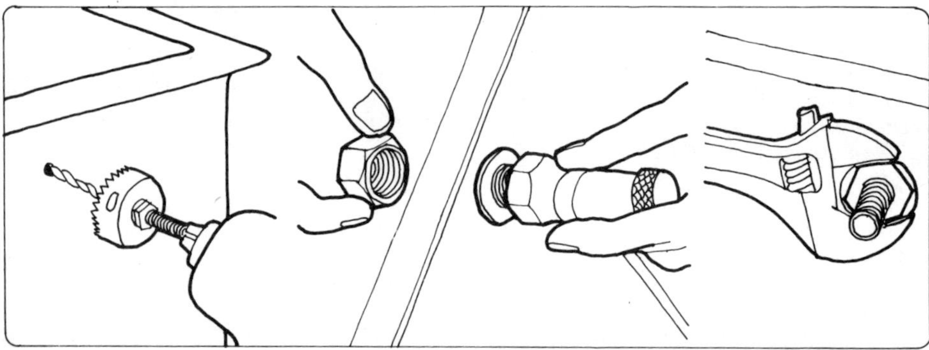

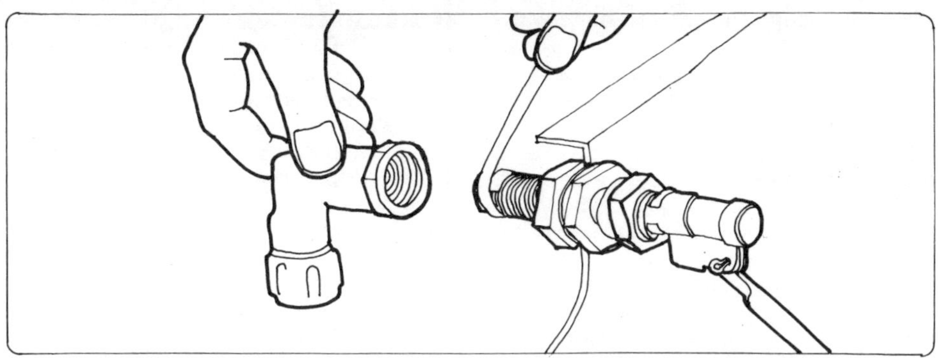

9.15 Teflon tape is wound around the float valve shank & a female iron to ½" compression elbow is fitted which will accept the cold water supply.

filled. You might therefore select an electronic controller which can turn on two pumps whenever the circuit needs to be refilled, but which cuts out one pump after a time interval corresponding to the filling time. This would reduce the amount of electricity you burn in catching the sun.

Another way of reducing the amount of pumping energy you use is to manually prevent the collectors from draining at times when there is no danger of frost occurring. This can be achieved quite simply by installing a check valve on the line between the pump and the bottom of the collector. This alone would prevent the collectors from ever draining down. A loop round the check valve is therefore required, with a gate valve in the middle. If the valve is closed, the collectors will remain filled. When frosts are possible, the valve should be left open so that the collector will automatically drain whenever the pump switches off.

Before filling, the system should be flushed out as described above and leak tested as described below.

Electronic Pump Control

There is a great variety of differential temperature pump controllers on the market and it is very difficult to give generalised directions for their installation. In all cases study the supplier's instructions. There are though, a few points worth bearing in mind while doing the plumbing work.

The controller should be located where it is easily seen by the householder, but not where it would be in danger of getting wet while draining or filling the system or in the case of leaks from the tank or pipe joints.

The sensors work best when they are in direct contact with the fluid, but it is usually adequate, and certainly easier to clamp them, or stick with epoxy resin, to the metallic wall of the pipe or tank. The collector temperature sensor is best placed on the rear of the absorber plate, but on the top outlet pipe, an inch from the casing, is a satisfactory compromise. The tank sensor should be filled midway between the two pipe connections that lead to the collectors. Make sure that it is connected to the wall of the tank itself, and not to a metallic cover surrounding a factory insulated tank.

The Tank Connections

The cylinder is filled by a line to the city water supply. The cold water should be delivered to the bottom of the cylinder, but often the supply line connects to the top of the tank, and a dip tube channels it below. A gate valve must be fitted to the supply line to enable draining of the system. Some local codes will also require a back-flow preventor to be installed on the supply line. This is a worthwhile precaution to prevent contamination of the supply water, particularly if you are using a toxic fluid in the collector loop. Although the function is similar, you should not install a check valve in the place of a back-flow preventor. There should be gate valves on each side of the back-flow preventor and perhaps also a strainer upstream of it.

If a two-tank system is being installed, the auxiliary tank is filled by the hot draw-off from the top of the solar cylinder. It is however normal practice to assemble gate valves and a branch from the supply line, in order that the auxiliary tank can continue to fill and supply hot water even if the solar tank has been completely drained. This arrangement is illustrated in figure 9.7.

The hot water draw-off which connects the cylinder to the faucet or shower, is sometimes led to a mixing valve which can blend hot water with cold, which it draws from the supply line to which it is connected. Mixing takes place whenever the temperature of the hot water exceeds a pre-set maximum. This layout is shown in figure 9.7. The mixer is a safety device which will prevent scalding from occuring when the water is unexpectedly hot.

Each cylinder must be fitted with a temperature and pressure relief valve at the top, and a drain valve at the bottom. The relief valve should have a pipe connection which terminates over a drain to dump any discharges. The valve is usually set to blow at 125 p.s.i. or 210 degrees F.

Checking for Leaks

A professional will probably use an air compressor to check for leaks. The do-it-yourself installer will probably have to fill the system then check carefully each joint. The system should be filled to one and a half times the intended operating pressures and left this way for an hour or two. If the testing pressure exceeds the maximum ratings of the expansion vessel or any other equipment, remove these and plug their connections before the test.

If any leaks show up, and they usually do, they will have to be fixed by tightening their nuts, or draining and resoldering, as appropriate.

Insulation

Pipe insulation is sold in the form of a sleeve which has to be slipped over the pipes before they are soldered together, or as a slit sleeve, or in preformed halves which can be fitted after the plumbing installation is complete. The latter has the advantage of allowing easy access to all joints during the leak testing procedure. The pipe insulation should have a minimum value of R4 which means using a thickness of at least 1" to 1½" of foam rubber or fibreglass, or ¾" of urethane or isocyanurate. Foam rubber sleeves should be butt jointed and glued together

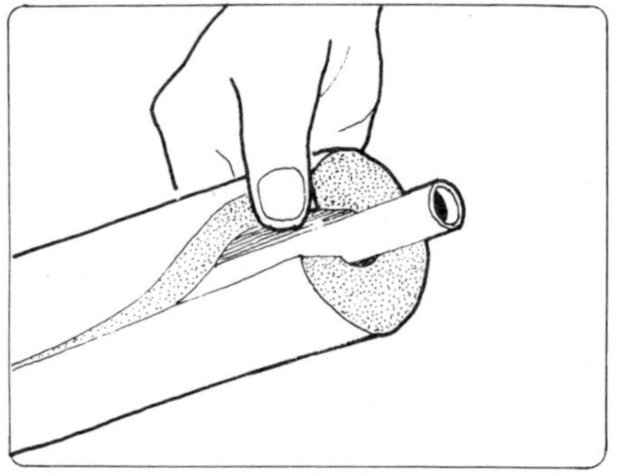

9.16 Pipe insulation requiring taped joints

in accordance with manufacturers' instructions. Preformed halves should be taped along the joints. Where it is installed externally, waterproofing and ultraviolet light protection is required. A manufacturer approved paint or plastic jacket is fitted over preformed halves.

Be sure to insulate around fittings and valves as gaps can loose a lot of heat. An exception to this is the cap of the air vent which must not be covered and the motor of the pump which would have a short life if kept operating at high temperatures.

The tank should have about three times as much insulation as the pipes to give it a minimum value of R11. You can buy tanks which are factory insulated to this level or you can box it with insulation boards held together with duct tape. Once again stuff up all gaps, particularly where pipes connect with the tank. If you are

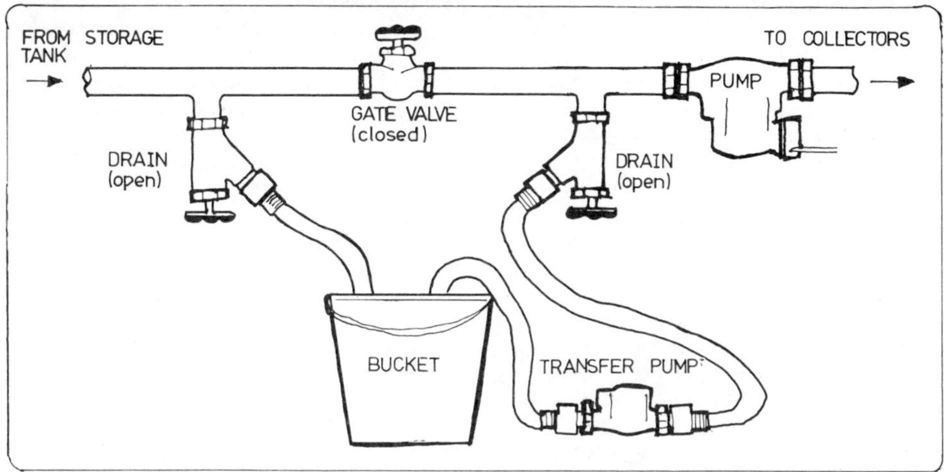

9.17 Filling the system

insulating an oil gas-fired tank, keep the insulation away from the burner to avoid fire hazards.

Flushing and Filling the System

Before filling the system must be flushed out to remove any debris which may have entered the pipes during installation.

First close the gate valve which sits between the two drain valves. The drain valve nearest the cylinder should then be connected with a length of hose to discharge over a drain or large bucket. Open both drain valves.

Clean water should now be forced in through the drain valve which is nearest to the pump, using a hose connection either to the main cold water supply line, or to a high head, self-priming pump which draws water from a barrel or large bucket. (The high head pump is the type you might use for bailing out a waterlogged basement.) Allow water to flush through the system for quarter of an hour, then stop and clean out the strainer in front of the solar circuit's pump.

Note If you are going to use silicone heat transfer fluids, or hydrocarbon oils, do not flush the system with water. Follow the manufacturer's directions.

Filling is a similar procedure. If you are using something other than water for the heat transfer fluid however, you cannot use the water supply pressure for filling. Use the high head pump and a large container filled with the desired solution. This time, the drain valve nearest the tank should be connected not to the drain, but to the same container from which the high head pump is drawing fluid. It will therefore be possible to circulate round the system, by-passing the closed gate valve which separates the two drain valves. This should continue for some twenty minutes to allow air to be purged from the system. After this, slowly close the drain valve through which fluid is returning to the container; i.e. the valve closest to the storage cylinder. Allow pressure to build up to the desired level before closing the second drain valve and turning off the high head pump. To prevent air infiltration the pressure should be at least 5 p.s.i. at the top of the circuit. If the pressure gauge is at the bottom of the pipework, it will also be registering the pressure due to the static head of water above it. The pressure at the bottom of a circuit therefore should be higher; e.g. about 15 p.s.i. at the bottom of a system with an overall height of 20 feet. Before you decide what pressure to aim for, make sure it is within the limits set by the manufacturers of valves and equipment you have installed.

Finally, do not forget to open the gate valve between the two drain valves, so that circulation will be possible when the solar system pump is activated.

The technique described above has been recommended in publications from the US Department of Housing and Urban Development and has proven practical. If though you find difficulty in getting hold of a high head pump (it is too expensive to buy for a single application), you might try filling from the highest point by disconnecting the vent and pouring fluid in manually. Pouring of course will not produce the desired pressure. A simple way of building up the pressure is to use a small manual garden sprayer for squeezing in the last gallon or two. If you use this technique, it will be necessary to add a gate valve between the tee fitting and the air vent seating, in order to seal off the pressurized circuit whilst removing the filling hose connector and replacing the air vent. When filling is complete, the gate valve must be left open in order that the air vent can operate.

For future maintenance purposes it is important that you make out a label listing the heat transfer fluid used, the date of filling, the desired operating pressure, and the manufacturer's maintenance recommendations. This should be hung round the pressure gauge.

Remember to check the pressure level during the first few days, particularly after the first sunny day, as there will be more air released through the air vent, and this may have reduced the pressure enough to make recharging desirable.

10

COLLECTOR CONNECTION PATTERNS

The pattern in which collectors are plumbed together has considerable effect upon their performance. Ideally, we want water to flow through the different collectors, and through the different waterways in each collector, at an even velocity. If the flow through one particular area is comparatively slow, a hot spot will appear because the heat is not being transported away so rapidly as in the surroundings. Hot spots cause greater heat loss which results in lower operating efficiency.

This chapter explains the factors which affect the flow of water through a solar collector and provides a guide to selecting the best connection pattern for different types of system.

Frictional Losses

The single most important factor in circuit design which affects the flow rate is frictional losses. The resistance offered by the pipe walls, especially at bends and fittings, tends to reduce the rate of flow. The frictional losses are assessed by measuring the length and diameter of pipework, and counting the number of fittings and junctions. Plumbers use tables which equate the friction due to different types of fittings with a length of straight pipe of equivalent resistance.

In order to achieve equal flow rates across all the waterways in a solar collector, we attempt to ensure that the frictional losses on each route through the collector are equal. This can be done simply by measuring the length of pipe comprising each alternative route.

Fig. 10.1 shows the kind of layout which is usually adopted. By placing the inlet and outlet at diagonally opposite corners the frictional losses will be equal along each route through the collector.

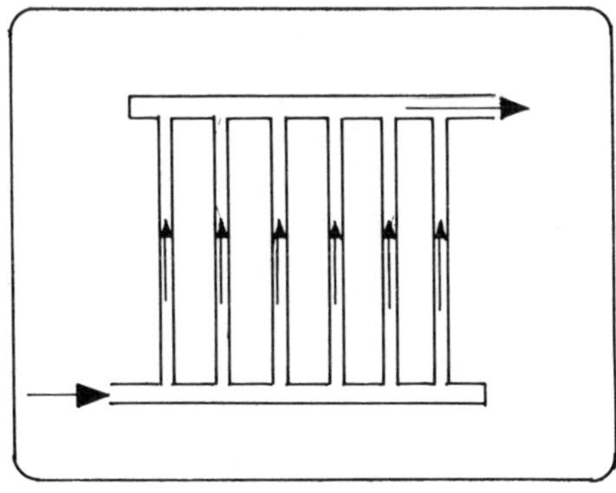

10.1 Collector connected at diagonally opposite ends to equalise frictional resistance along each path.

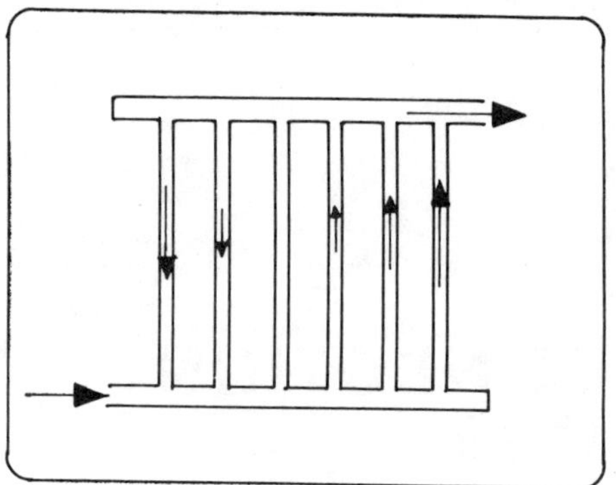

10.2 Uneven flow pattern resulting from Bernoulli effect

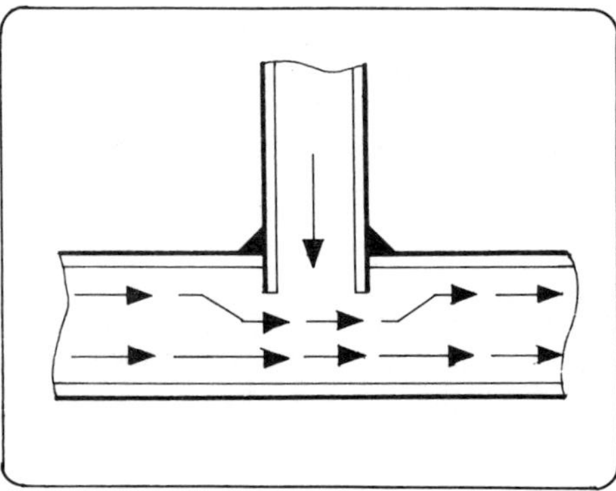

10.3 Bernoulli effect: Where riser projects into the header pipe, a negative pressure can be created at the mouth of the riser due to the increased velocity of flow through the constricted header.

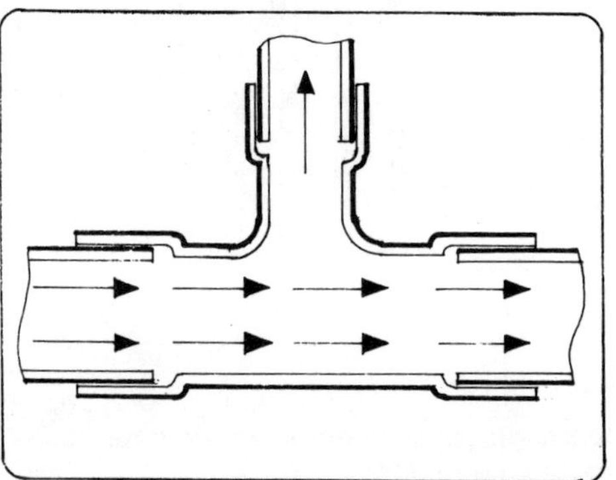

10.4 Where a flush tee fitting is used to connect riser and header, there is no problem with Bernoulli effects

Bernoulli Effects

Some researchers however have measured the flow which occurs through commercial solar collectors and discovered that the flow pattern is sometimes quite different from what we would have expected. With certain collectors there has been no flow in the central risers. The most rapid flow has been in the risers nearest the outlet while flow has actually been downwards in the riser nearest the inlet (see fig 10.2). These strange results can be explained in terms of the Bernoulli effect. They are only likely to be found in collectors whose construction method results in the end of the riser tube projecting into the header pipe as in fig. 10.3. This constitutes the channel through which water can flow. As a result the velocity increases and this in turn results in a drop in pressure which can be strong enough to suck water down the riser. This effect is unlikely to be noticeable when a flush tee fitting which does not restrict flow is used (fig. 10.4). Other ways of preventing the Bernoulli effect from causing uneven flow patterns, is to use header pipes which are relatively much larger in diameter than the risers. Alternatively the risers should be made much longer than the headers. Both means tend to increase the importance of frictional losses to the point where they outweigh the pressure drops caused by Bernoulli effects. It is also the case that the Bernoulli effects will only cause problems at relatively high velocities. Thus they are of no concern in thermosyphon systems.

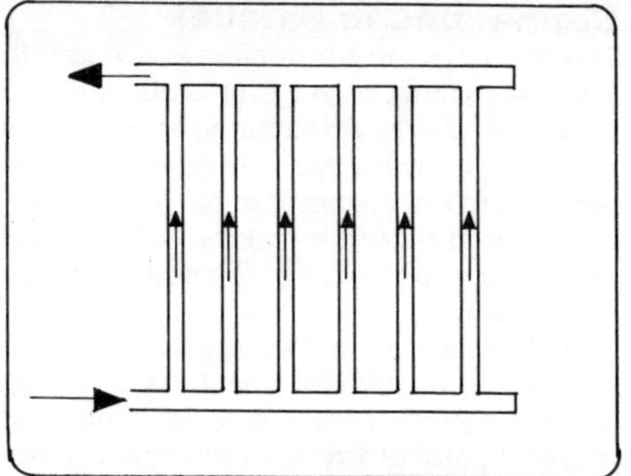

10.5 Where the collector construction gives rise to problems with Bernoulli effects, these can be corrected by connecting the inlet and outlet to the same side of the collector

Problems due to Bernoulli effects can be ironed out at the system design stage by connecting inlet and outlet pipes to the same side of the collector. This tends to even out the pressure drops across the risers and thus leads to fairly equal velocities (fig. 10.5) except where the risers present a proportionately large frictional resistance. The same principle is true for arrays of solar collectors as indicated in figs 10.6 and 10.7.

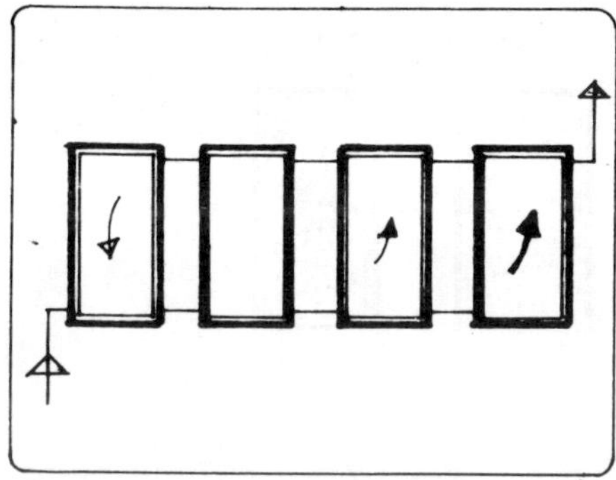

10.6 With diagonally opposite inlet and outlet on a parallel array, circulation through the collectors may be uneven when pumped if the collector encourages the Bernoullli effect

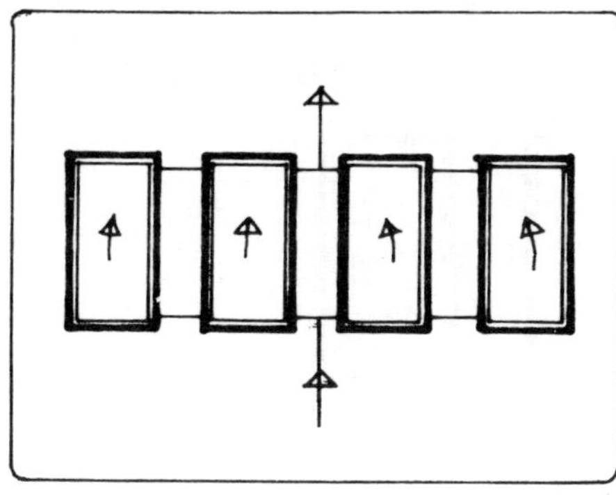

10.7 The Bernoulli effect in a pumped, parallel array of solar collectors, can be countered by locating inlet and outlet pipes centrally

Connecting in Series

With series connection, the outlet of one collector is fed to the inlet of another collector as in in fig 10.8. In this case there is no problem about different flow rates in different collectors. The flow rate is equal throughout. The frictional losses in a series connection are greater. Therefore more pumping power is required, or the flow rate will be comparatively less, the operating temperature higher and subsequently a drop in efficiency will be experienced.

Connecting in Parallel

The alternative to connecting in series, is in 'parallel'. In this case water is delivered to each collector by a header pipe which runs along their lower edge. The header can be external to the collector or an integral part of it. Heated water is withdrawn via a header along the upper edge of the collectors (as in fig. 10.9).

Friction losses are lower across collectors connected in parallel and thus pumping power requirements are lower. This fact makes a parallel arrangement seem somewhat preferable for thermosyphoning systems where the circulation force is very weak.

The disadvantage of parallel connections is the increased difficulty of achieving a

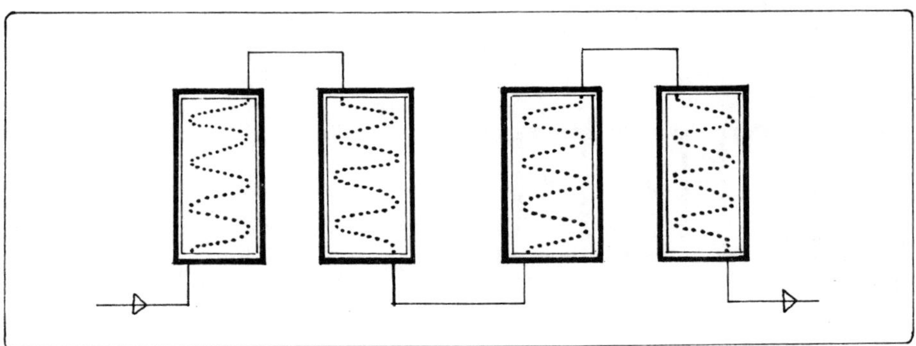

10.8 Collectors connected in series

10.9 Collectors connected in parallel

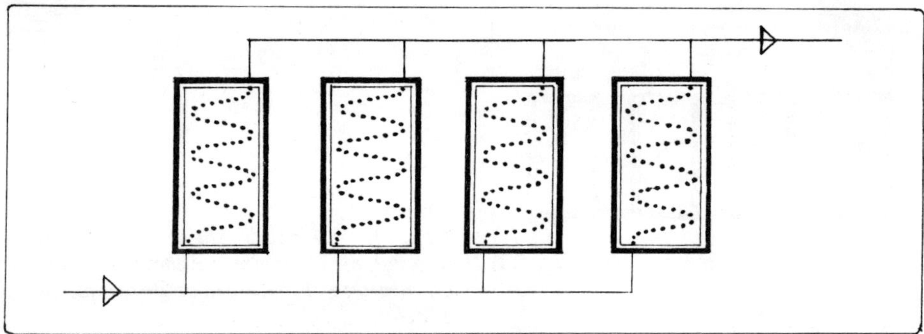

balanced flow. This is particularly so if the collector construction is such that Bernoulli effects might become significant.

In cases where the header pipes cannot be installed with a reverse return layout to ensure that the length of piping encoutered on the path through each collector is equal, balancing valves may have to be installed. These are particularly useful on large arrays of solar collectors enabling the friction losses along each route to be modified.

Examples

Different systems therefore require different solutions. Similarly, the collector construction will have great effect upon the internal flow pattern and hence upon the choice of connection layout. It is difficult therefore to make generalised recommendations, but the following diagrams will at least indicate the pros and cons of various alternatives.

Collectors in Thermosyphoning Systems

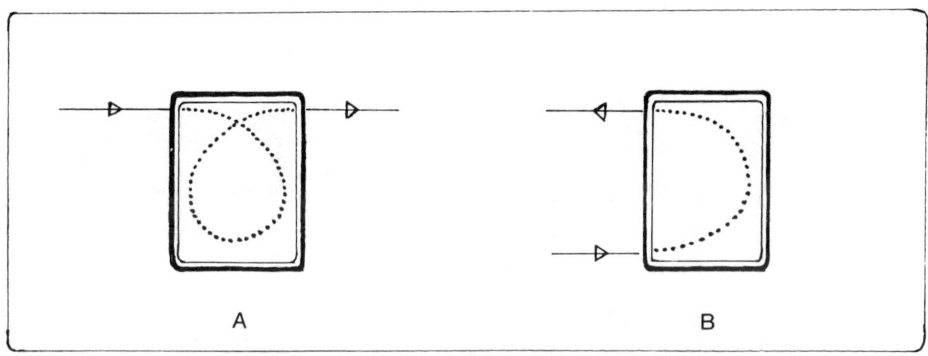

10.10 B is better than A because it introduces unheated water at the bottom of the collector. Introducing cold water at the top disturbs the convective currents.

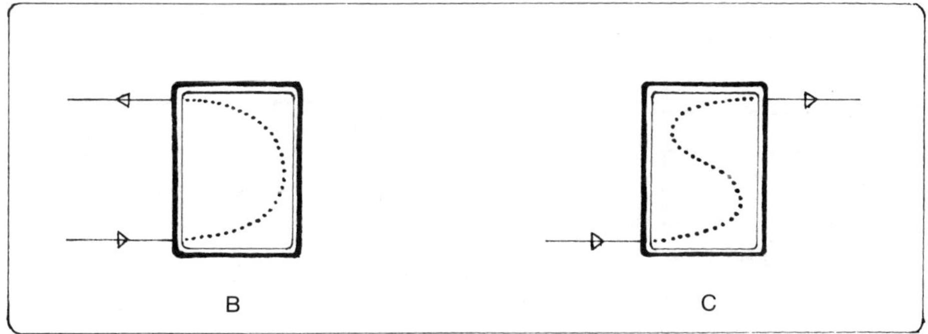

10.11 C is better than B because it will discourage the formation of 'dead' corners where no flow occurs.

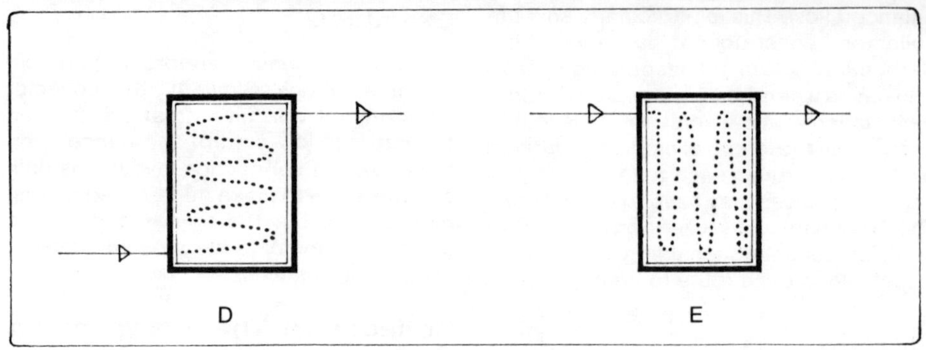

10.12 With serpentine tube waterways, D is best because it provides a continuous upward gradient to assist convective flow. E would also be impossible to drain down.

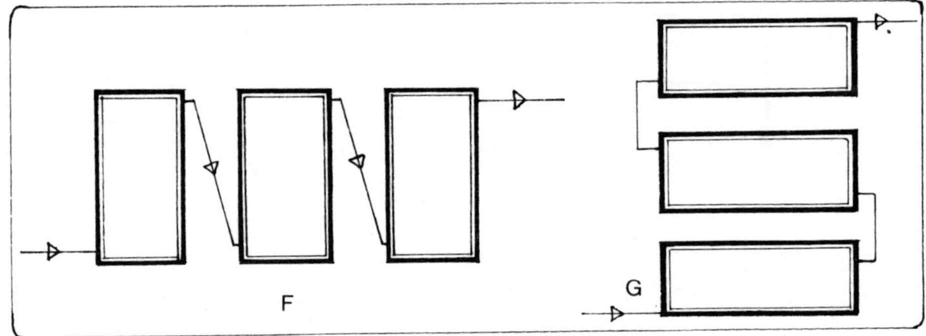

10.13 G is better than F because it allows the circulating fluid to rise continuously without forcing the flow downwards as in F.

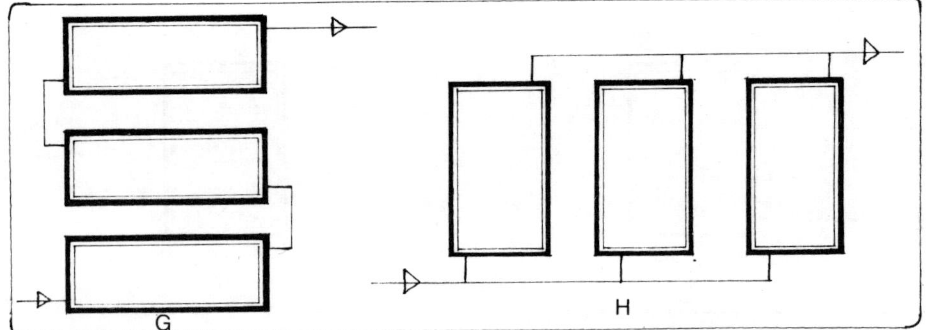

10.14 H is better than G because it offers less frictional resistance to flow.

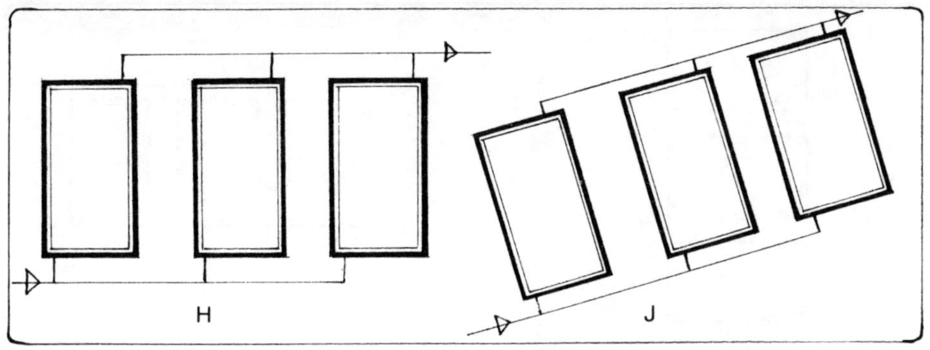

10.15 J is better than H because by giving the horizontal headers a slight tilt towards the outlet, it reduces the resistance to the convective force. The water will therefore circulate faster and more energy will be collected.

Collectors in Pumped Systems

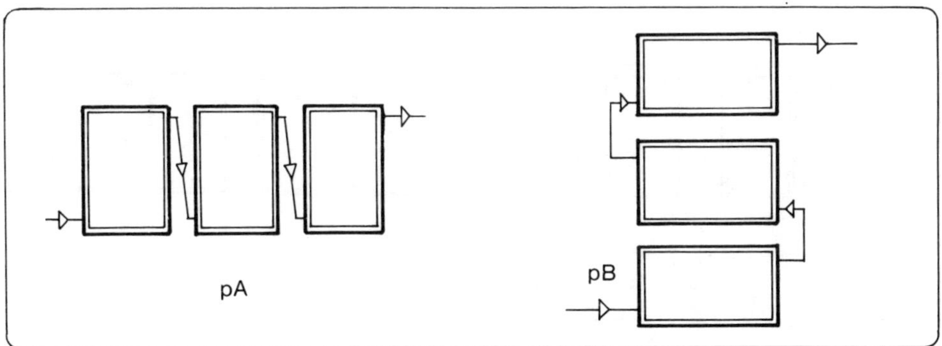

10.16 pB is preferable because there is a danger of air blocks occurring where the outlets from the pA collectors bend downwards.

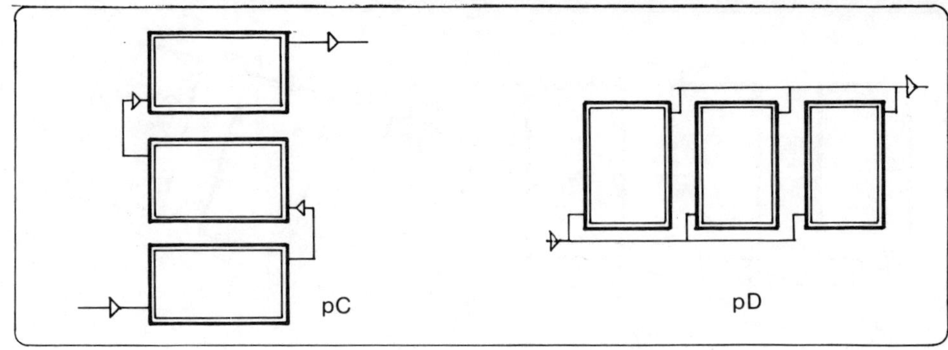

10.17 Given the same flow rates these systems should perform equally.

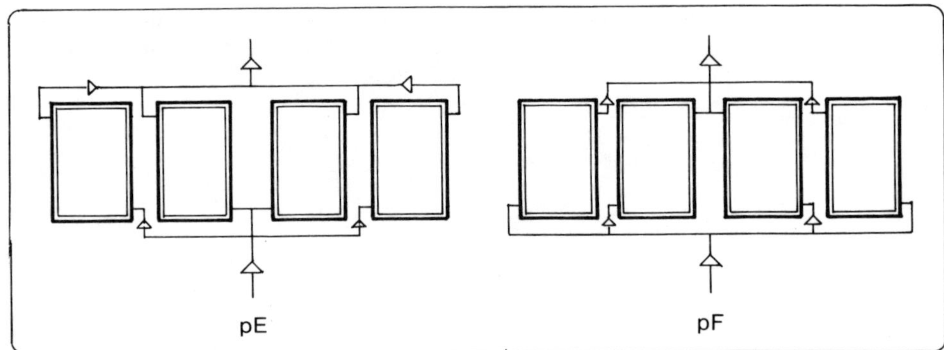

10.18 pF is a little better than pE because it has a smaller length of exposed outlet from which to lose heat.

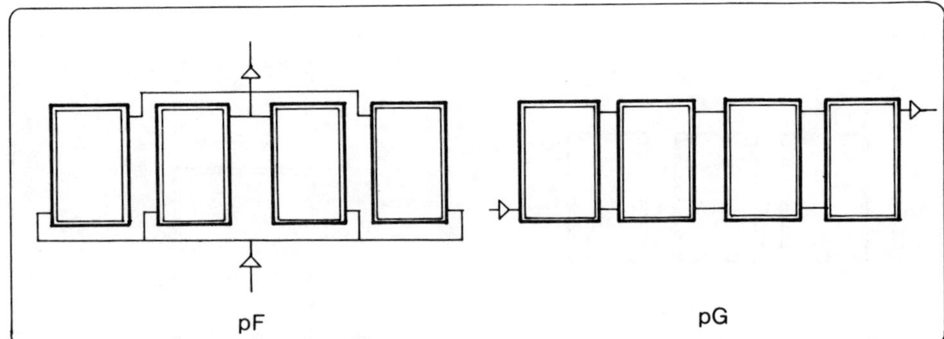

10.19 pG loses less heat due to having its header pipes inside the collector casing. It also has fewer bends thus reducing friction losses.

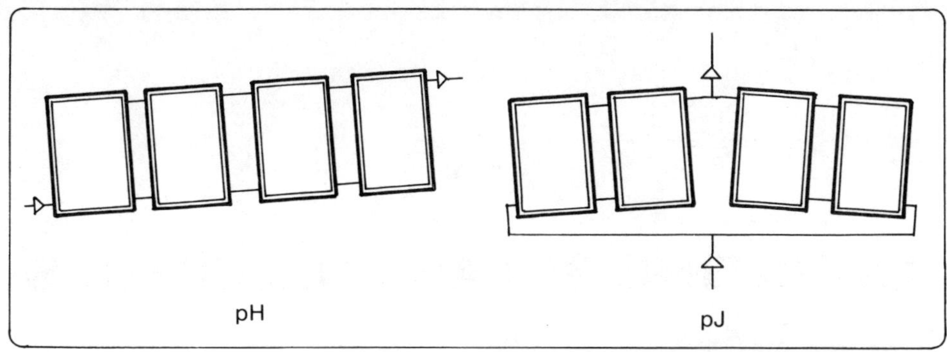

10.20 pH and pJ both have slight tilts in their headers to ensure that trapped air will be flushed out, and draindown will be easy.

11

MOUNTING SOLAR COLLECTORS

Location

The primary requirement in choosing a location for the collectors is that it should receive as much radiation as possible. They should face no more than 30 degrees away from due south. If you are not sure exactly which direction south is, find out the times of sunrise and sunset, and work out the hour midway between them. This is the solar noon and at that time of day the sun itself will show you due south. Avoid any obstructions, like trees or buildings, which might cast shadows for a large part of the day. Raised areas, like roof tops are usually less overshadowed than walls or gardens.

Having picked out the sunniest areas one should then consider the house's plumbing layout. The longer the connecting pipes between collectors and hot water faucets, the more heat will be lost on the way. Ideally collectors, storage tanks and faucets will be close together.

Finally, in remote locations the type of circulation desired may be important. If there is no electricity supply to run a pump, the system must be planned to operate with gravity circulation. As explained in chapter 5, this means that the collectors must be mounted below the solar storage tank. Single story extensions, balconies, walls and ground level locations may be most appropriate in such cases.

Appearance of Collectors

Solar collectors need not greatly affect the appearance of a building. In some cases they can even enhance it. They have become an eyesore in some instances where frames are built, jutting out at odd angles from the roof or walls with no relationship to the existing elements of the buildings. Where possible the collectors should be installed so that they line up with any strong lines formed by windows, doors or projections in the building facade.

Mounting on Pitched Roofs

Solar collectors can be inset mounted so as to replace a conventional roof covering or can be surface mounted above the existing tiles or shingles.

When the collector is to be an integral part of the roof, the existing tiles must be removed and a supporting framework built on the plywood sheathing above the rafters. The framework is flashed with zinc at

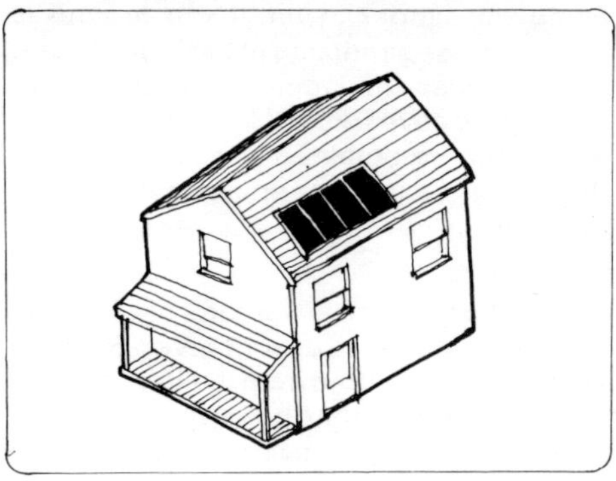

11.1 and 11.2 Collectors mounted on pitched roofs. Location of collectors will depend on form and orientation of the house.

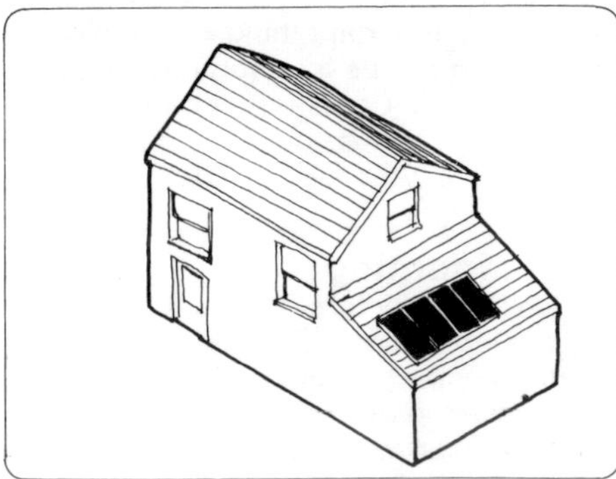

the edges and after the absorber plates are plumbed in, glazing adds the final weatherproofing.

Surface mounting the collectors can be a much quicker operation. In this case metal angles are bolted through the roof and the collectors secured, top and bottom, between them. With this type of mounting, the collectors should have a durable casing such as fiberglass or aluminium. A wood cased collector could be surface mounted but a weather resistant timber must be used or it will be necessary to repaint the casing about every two years in order to maintain a weatherproof finish.

Safety and Access

The majority of accidents in the building trade occur when people are working with ladders. This does not mean you have to go rushing off to your local builder when it comes to roof work. But it does mean you ought to develop the necessary confidence by careful consideration of how to arrange access to your particular roof.

Extension ladders or scaffolding are the obvious means of access and climbing boards, or roof ladders, which have a curved lip at the top for hooking over the roof ridge, are necessary for actual working on the roof.

Numbers

Two people is probably the best number for working on the roof. More than that can be distractive and may get in each other's way.

Ladders

If you are using ladders then it is best to have two sets, one on each side of the area being worked on. The ladders should be pitched against the building so that the base is away from the wall by a quarter of the height it is reaching. The ladder should be secured at the top and the bottom. If there are no existing objects against which it can be secured, screw a ring bolt into the woodwork under the eaves, and drive stakes into the ground to provide anchors for

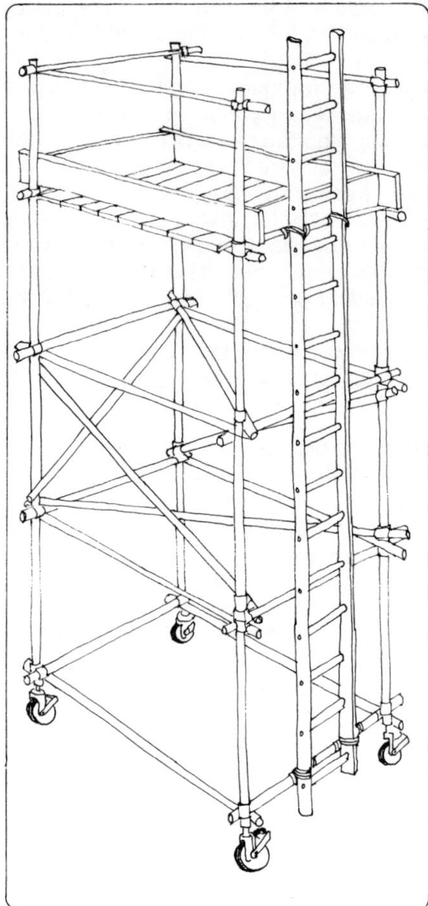

11.3 Scaffolding tower—note ladder tied top and bottom.

tieing the ladder down. If the ground is concrete, use sandbags at the ladder base.

Do not lean the ladder on plastic gutter pipes. If the eaves project by a large amount, a stay can be obtained which hooks over the top of the ladder and holds it out from the wall.

Scaffold

Scaffolding has the advantage of providing a working platform at roof level. Scaffold sections can be hired from builders and are easily erected as they come in component form and each one simply slots into the next. They usually have wheels on their base and these should be locked before commencing work on the tower. The tower should also be tied to the building you are working on.

The most important thing is your mental attitude. Do not be in a hurry. Remember where you are. Concentrate on the job. If you find that impossible, then maybe you should call in your local builder.

Inset Mounting

Setting the collectors into the roof is probably the most economic method of mounting for anyone building their own system. All exposed parts of the framework encasing the absorber plates are covered with metal flashing hence the cheapest softwood can be used.

The major disadvantage of inset mounting, is the increased time spent working on the roof top.

Inset Mounting Procedure

The first step is to mark with chalk the intended location of the collectors. General hints about choosing a location are given above. Considering the practical point of view, installation will be easier if they are sited away from the roof edges to avoid disturbing any existing flashing. Access for making pipe connections inside the roof will be facilitated if the collectors are a few feet above the eaves. Remember however that you should aim to have space inside the roof to install an air vent on the outlet

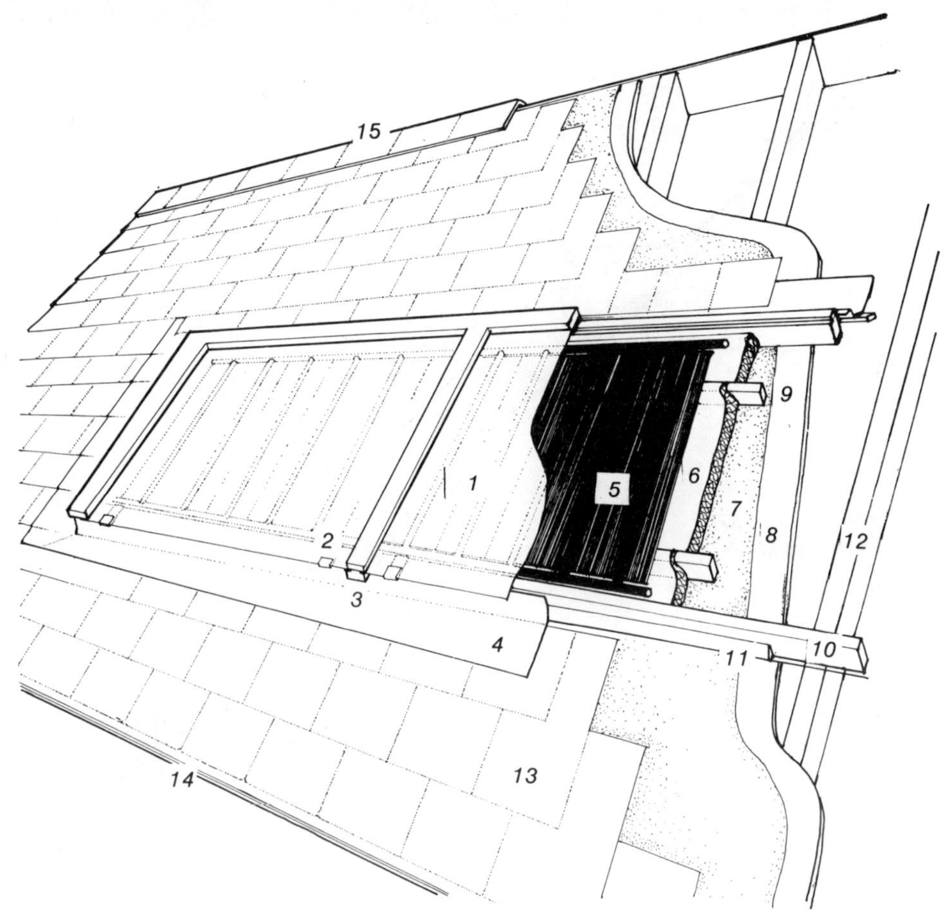

11.4 Inset mounted solar collector
Key
1. Glass
2. Metal glazing clip
3. Glazing box
4. Flashing
5. Absorber
6. Insulation
7. Roofing felt
8. Plywood sheathing
9. Absorber supports
10. Casing member
11. Triangular fillet
12. Rafter
13. Asphalt shingles
14. Eaves
15. Ridge

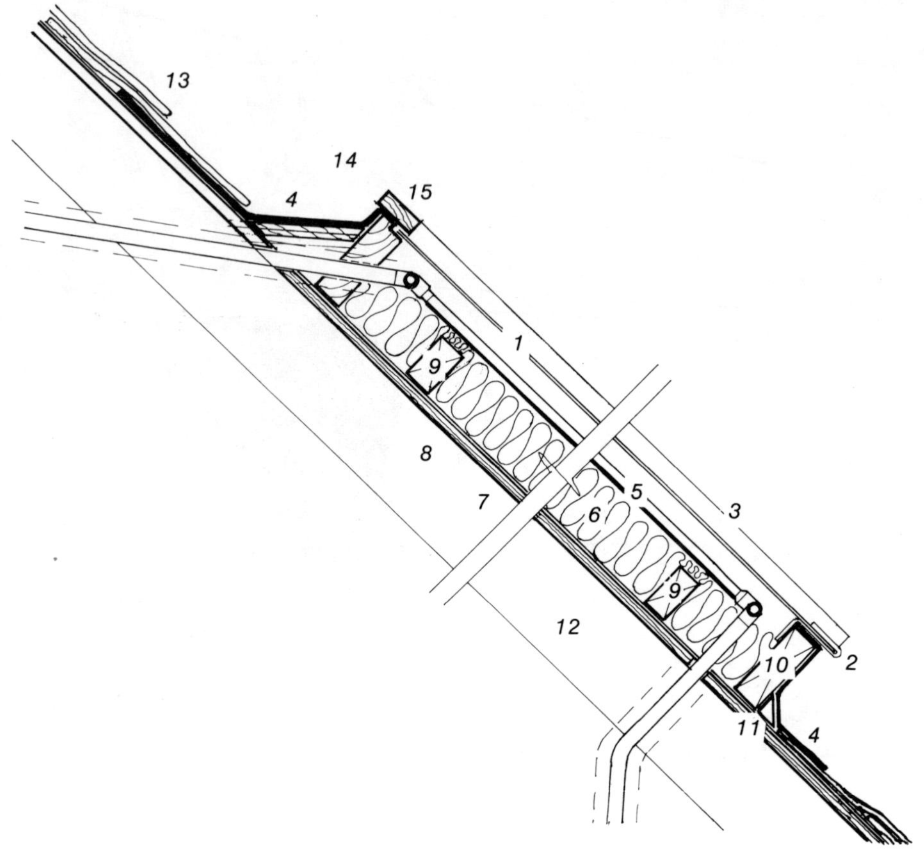

11.5 Section through inset solar collector

Key

1. Glass
2. Metal glazing clip
3. Glazing box
4. Flashing
5. Absorber plate
6. Insulation
7. Roofing felt
8. Plywood sheathing
9. Absorber supports
10. Casing member
11. Tringular fillet
12. Rafter
13. Asphalt shingles
14. Gutter board
15. Capping piece

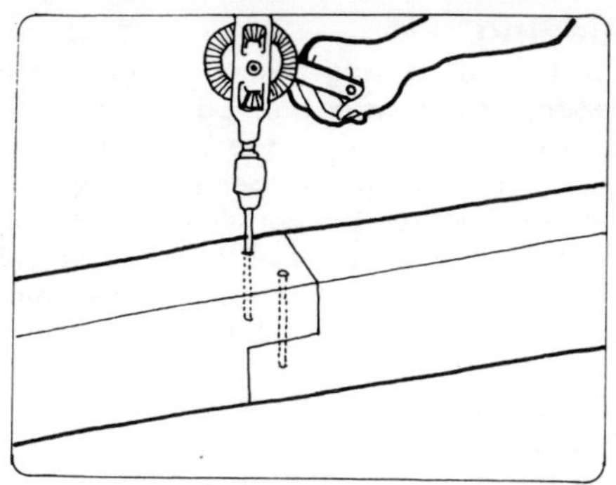

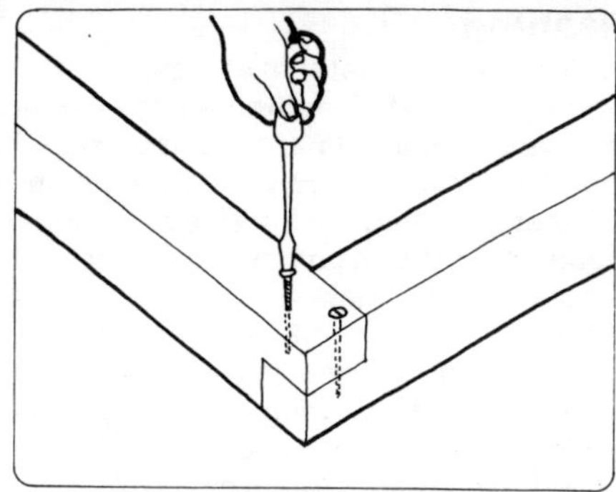

11.6 and 11.7 Casing framework is cut to size using lap joints

pipe above the top of the collectors.

When mounting the collectors in a horizontal bank, the short dimension will be equal to the height of the absorber plate plus 1" for tolerance, plus 4" for the casing frame members. The length will be the total length of all the absorbers plus an allowance of between them for fittings and at each extremity, plus 4" for the framework.

Casing

There are three basic approaches to casing site-built collectors. The simplest method is to use 2" × 6" softwood for the frame and aluminium angle to secure the glazing. This method is described in Chapter 12 and has the advantage that no machining of wood is required. An alternative to this is to make up a casing framework of 2" × 4" members, (figs 11.6, 11.7) to which separate glazing bars are attached (figs 11.10, 11.11). A third method (shown in the key drawings 11.4, 11.5) requires the formation of a very shallow (3/16" × 3/4") rebate in the top and side rails of the casing. This rebate provides a seating for the glass, and the bottom rail is planed down to allow the glazing to overlap.

Within and connected to the casing framework two 2" × 3" members provide support for the absorber plate. Above the top rail of the casing, two tapering gutter boards should be fixed so that they meet on the centre line of the collector array, and fall to each side to encourage water runoff. On the outer edge of the two sides and bottom of the casing, triangular wood fillets should be fixed to form a seating for the flashing.

With the wood framework complete, 3" of insulation (fibreglass) should be fitted prior to installing the absorbers.

Installing Absorbers

The absorber plates should now be laid upon the blocks in the casing. When working on a scaffold tower, they can be hauled up with ropes. Alternatively two people should take an end each and lift them between the two ladders. Once installed they should be connected in the most appropriate pattern shown in chapter 10.

Care must be taken with the hot water outlet where it goes through the roof to join the flow pipe leading to the solar storage tank. This outlet pipe should rise until it meets the air vent. If you simply took an elbow out of the absorber and led the pipe downwards into the attic, there would be a danger of an airlock forming at this point and reducing or blocking the flow rate.

It will be necessary therefore, except on steeply pitched roofs, to drill through the underside of the casing framework in order to allow this slight upward gradient in the outlet pipe.

On shallow pitched roofs this may be impossible and space inside the casing should be allowed for an automatic air vent to be fitted on the outlet at its highest point.

Flashing

A roll of flashing material at least 10″ wide is now required to weatherseal the casing. Lead and aluminium are the materials in common use for this purpose. Aluminium is cheaper than lead but not so easily formed round bends. Lead comes in several thicknesses and for this application 4 or 5 lbs per sq. ft. is recommended. A much cheaper, but also shorter lived material would be aluminium faced bitumin roll. This is easily moulded round corners and is self-adhesive.

The flashing should be nailed with galvanised clout nails to the upper surface along the lower edge of the frame. This sheet should project at the edges where it will be overlapped by the side flashing. Flashing on the upper edge is fixed last. This forms a kind of gutter for the roof above. If it is a large area of roof the gutter must be deeper. In such cases a wider strip must be used and this should extend over the glazing bar. The upper strip of flashing must be overlapped by the layer of shingles above the collectors and the joint waterproofed with mastic.

Aluminium flashing should not be fixed in contact with, or in a position where it would receive the run-off water from any copper piping or sheet. Otherwise the flashing would corrode.

Glazing

The type of glazing system adopted will depend on the casing method, as described above. A glazing technique which does not require machining of timber is described in the next chapter. For people who possess a circular saw (bench saw), the machining of rebates either in the casing members themselves, or in separate glazing bars, should not prove to be difficult.

Where separate glazing bars are made, a deep rebate should be formed (¾" × ¾"), and the glass bedded in silicone mastic and secured with stop molding (see 11.10 and 11.11). In such an exposed position however, the molding (unless it is good

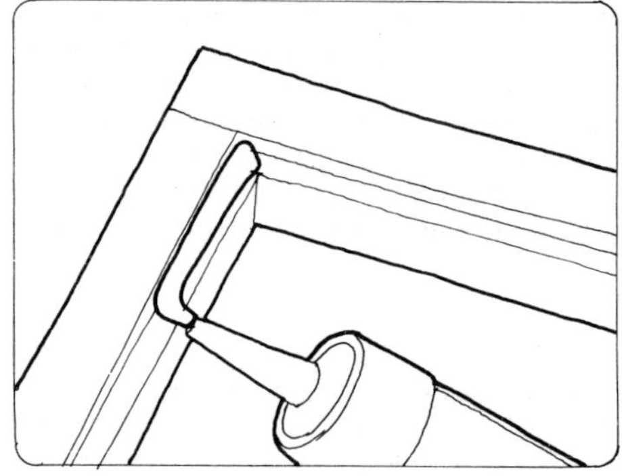

11.9 Sealing mastic for bedding glass

11.10 Fixing glazing bars in place

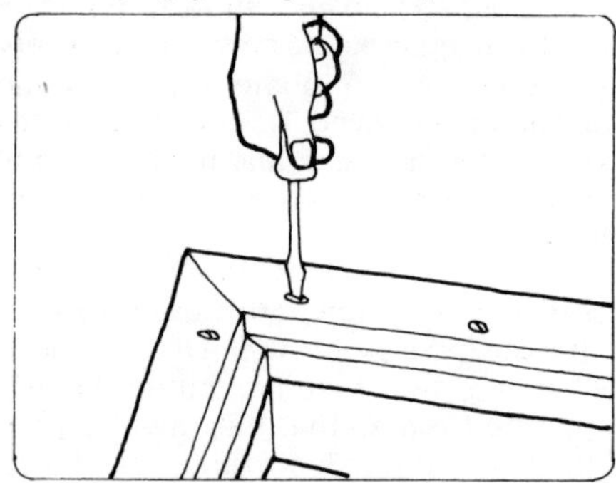

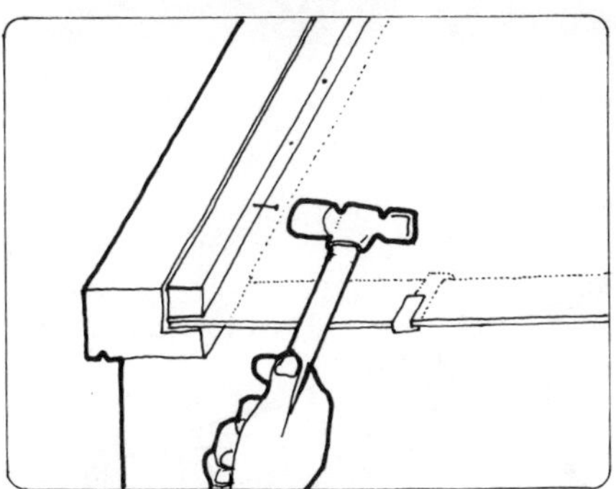

11.11 Nail molding to glazing bar after bedding in mastic

quality hardwood) may not last many years. A more satisfactory method is therefore to machine a shallow rebate in three sides of the casing frame (as described above), bed the rebate in mastic to receive the glass, fix the top edge of the flashing to the top of the casing frame, seal with mastic and cover with a 1″ × 2″ hardwood trim cap. The trim cap should be screwed down to the frame, the screw heads being countersunk and covered with mastic.

Along the lower edge, the glass projects over the frame to carry rainwater clear of the casing. It is held in place by metal glazing clips, nailed or screwed to the inside of the lower frame member before the glass is laid in place. A neoprene glazing strip or mastic should be run along the top of the lower casing frame to provide a draughtproof seal with the glass.

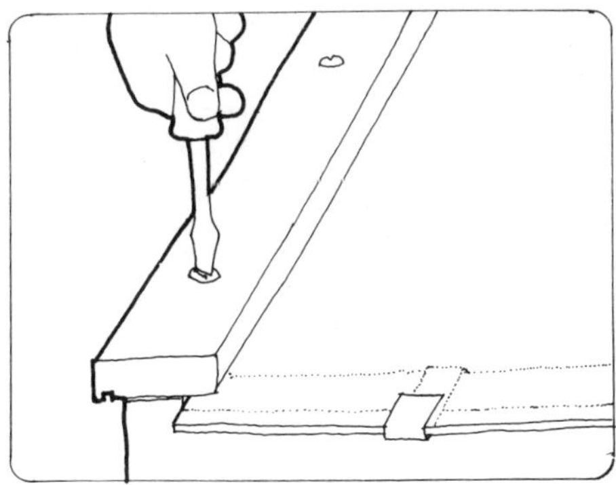

11.12 Trim cap

Surface Mounting

Manufacturers who produce solar collectors with durable casings often recommend a much quicker form of surface mounting. Many kits are fixed by bolting to a metal angle which is clamped above the roof by bolts which pass through the roof. Holes for the bolts are drilled and a weather-seal is achieved by packing the hole with silicone sealant.

When surface mounting collectors, it is important to realise that you are not concerned simply with preventing the collector slipping off the roof. The main threat is the wind, not gravity. In the most exposed conditions, where the collectors are mounted on the lee side of a gable roof, the upward acting suction force might be more than 20 lbs/sq. ft.

The collectors must therefore be fixed both top and bottom with at least four bolts or screws and these should be ⅝″ diameter. They should be fixed near each corner in order to prevent any rocking movement during windy weather.

11.13 Surface Mounted Collectors
Note the use of a roof flange where the pipe penetrates the roof. Notice also that the outlet pipe rises to an air vent, mounted inside the roof space, prior to descending to the storage tank.

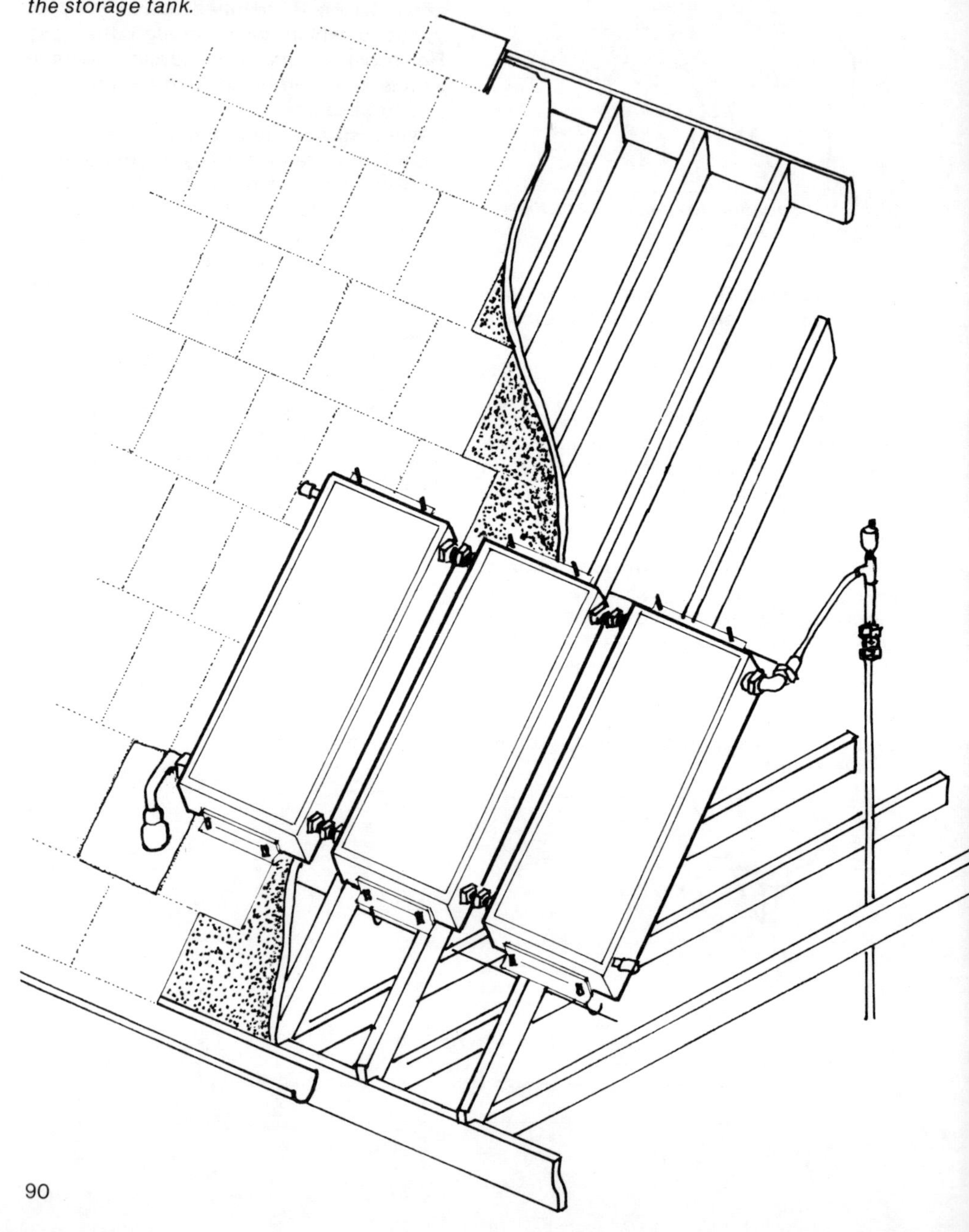

11.14 Mark intended positions of the solar collector on the roof using a chalk snap line and the locations of the bolt holes

11.15 Collectors must be raised above the roof surface to avoid trapping leaves and moisture which could damage roof surface. A 3" length of aluminium tube can be used for this purpose. To avoid damage to shingles the load should be spread on bearing plates.

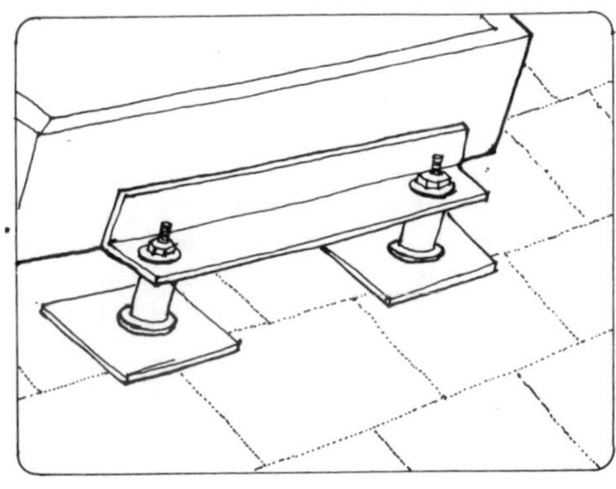

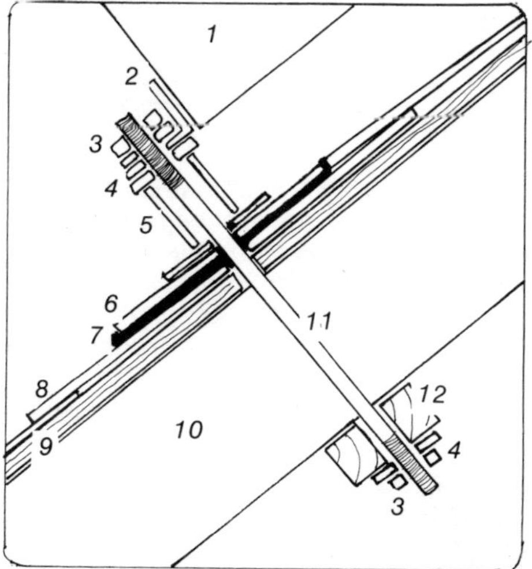

11.16 The Collectors are secured by bolting through to a length of 2" × 4" timber running behind the roof rafters.

Key

1. *Collector box*
2. *Aluminium angle*
3. *Nut*
4. *Washer*
5. *Aluminium spacer tube*
6. *Bearing plate*
7. *Mastic*
8. *Shingles*
9. *Plywood sheathing*
10. *Rafter*
11. *Threaded bar*
12. *4" × 2" bearer*

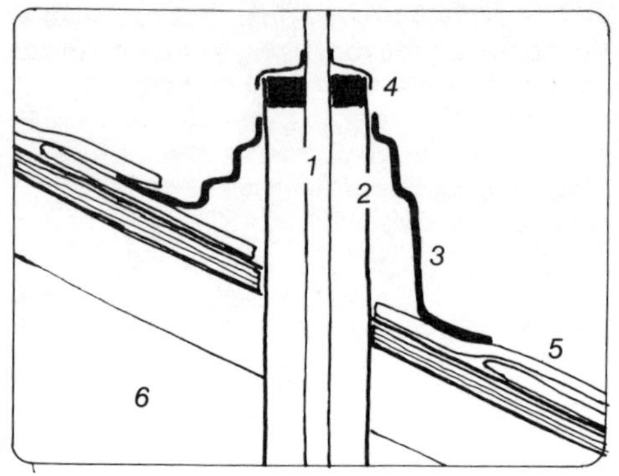

11.17 A roof flange is used to waterproof the hole where pipes pass through the roof surface. Flanges are available in a range of sizes, and have a neoprene collar which grips a PVC pipe sleeve, through which the copper pipe passes. The flange is sealed to shingles with roofing mastic.

1. *Copper pipe*
2. *PVC sleeve*
3. *Roof flange*
4. *Flexible cap*
5. *Shingles*
6. *Roof space*

11.18 Collectors are best connected with unions or compression fittings. This is because working close to the collector casing and the roof surface with a blow-torch can cause damage, particularly to the neoprene pipe gaskets which seal the joint between casing and pipe.

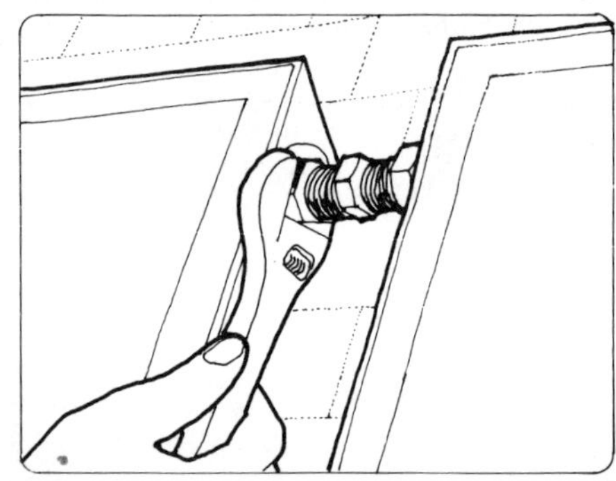

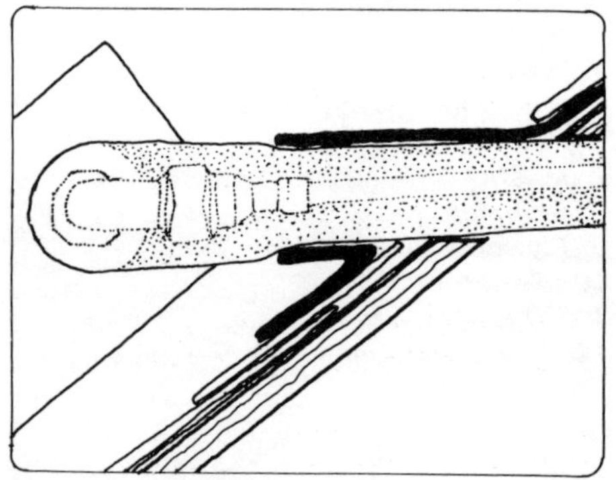

11.19 An alternative to the use of a PVC pipe sleeve is to have the roof flange grip a layer of rigid insulation around the copper pipe. Note that this outlet from the top of the solar collector has a slight gradient. It rises towards an air vent inside the roof space.

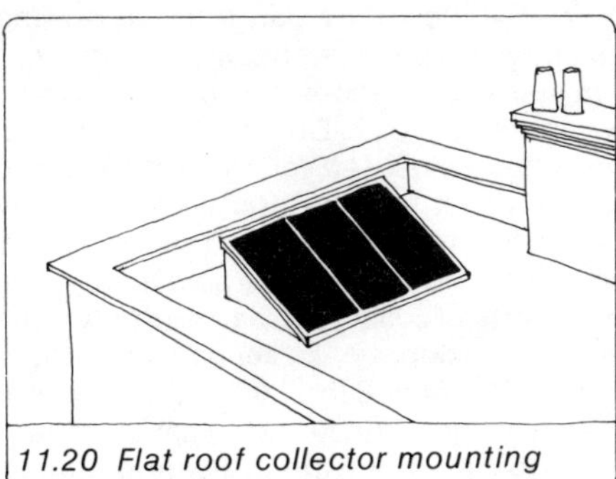

11.20 Flat roof collector mounting

Flat Roofs

On a flat roof a supporting framework is required in order to tilt the collectors towards south. Several companies sell metal frames to support their solar panels. A single triangulated frame can be constructed from wood. The wood should be treated with preservative and should not be laid directly on the roof surface in such a way that it might be unventilated. Thus the base plates of the structure should be fixed on spacer pads. These pads can be formed from roofing felt, nylon, lead or any material which will not corrode in wet conditions.

In choosing the collector location, factors such as overshadowing are of primary importance. But it would be wise where possible to avoid those areas of roof where puddles form after rain.

11.21 Wall mounted collectors integrated with the building facade

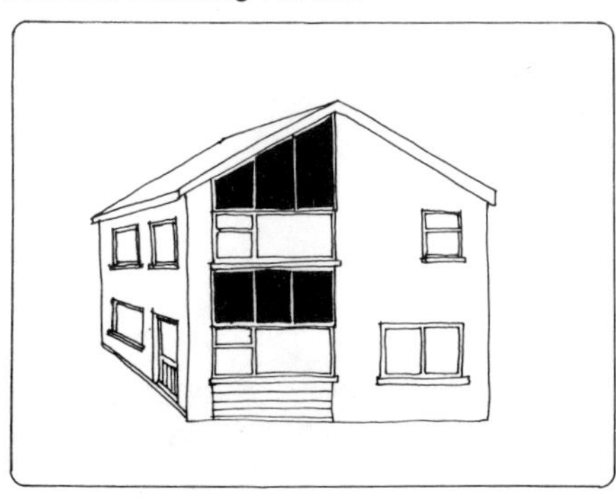

Vertical Mounting on Walls

In some situations a wall may be the only available location for solar collectors. This will undoubtedly have a greater effect on the appearance of the house. This need not be a detrimental effect. Indeed, they can add both to the appearance and the function of the building. Mounted over an entrance they can form a sheltering roof.

11.22 Wall mounted collectors fixed on a simple triangulated frame

Over a window they can form an awning which would provide shading in summer.

In these applications the collectors would be mounted at a tilt. Collectors have sometimes been mounted vertically. Such steep angles of tilt can perform well on sunny days in the middle of winter. Taken over the whole year however, you would need many more vertical collectors than 45 degree collectors in order to yield similar quantities of energy. Projecting the foot of the collector away from the wall by even a small amount can cause an improvement in performance.

One company uses ⅛" aluminium angle to wall mount their collectors. Another uses a triangular steel web to fix to a central pivot point on their collectors thereby allowing the angle of tilt to be altered.

12

MAKING YOUR OWN SOLAR COLLECTORS

A simple solar collector which can be made in a workshop at home, is the tube and sheet collector. The choice of tube configuration (either serpentine or parallel grid) relates to the type of system of which the collectors will form a part. For example, a parallel grid collector is not the best choice for a draindown system due to the risk of air-locks forming when the system is refilled. Where a gravity circulation (thermosyphoning system) is adopted, care should be taken with serpentine absorbers that the tubes provide minimum resistance to flow.

The most suitable material for tube and sheet absorbers is copper, although aluminium and steel can be used (see Chapter 4). Where copper is used, the bond between tube and sheet can be obtained by soldering, which will provide long lasting thermal contact. However, before the construction of the absorber plate is discussed in detail, consideration must be given to the overall dimensions and materials of the collector casing, and the final location of the panels. Thought should be given to the use of panels larger than 5ft x 2ft because they are difficult to handle on rooftops. However, slightly larger panels may be justified if, for example standard sized tempered glass from patio doors is available at a good price. Obviously, where a collector array is to be installed integral with the roof covering (see Chapter 11) component dimensions will vary as the construction of the casing takes place on the roof. In this chapter, however, the emphasis is on the workshop construction of a complete panel for installation above the roof covering.

Casing

Aluminium and fibreglass are the most common materials used for encasing commercial collectors due to their light weight and durability. Unless you have had experience in working with these materials, however, you will probably find wood easier to work with. Softwoods will require preservative treatment when the frame is first made up, and at regular intervals afterwards.

Tables 12.1 and 12.2 show the materials required to construct a wood casing for the absorber plate. Precise dimensions are not given, as these will vary according to your own situation. However, having determined the overall dimensions, the dimen-

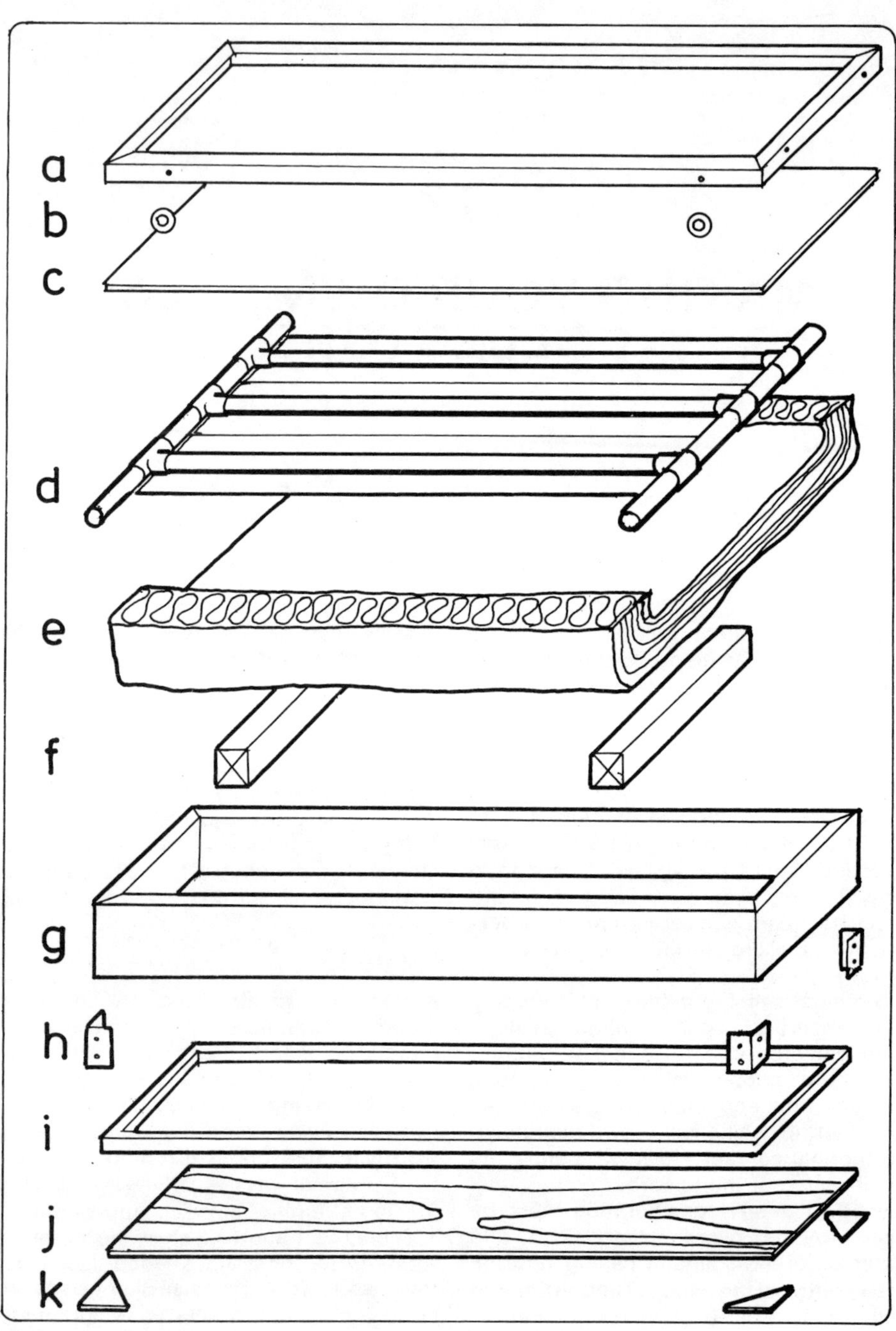

12.1 Exploded view of a solar collector

COMPONENTS

Ref.	Component	No. Reqd.	Material	Dimensions Inches (mm)
a	Glazing angle	2 2	Aluminium angle	
b	Glazing angle spacer	10	Washers	
c	Glazing	1	Glass	
d	Absorber plate	1	Copper Tube-and-Strip Absorber	
e	Insulation		Glass fibre duct insulation	
f	Absorber support	2	Softwood battens	
g	Side panels	2 2	Preferably redwood	
h	Corner angle	4	Aluminium angle	
i	Base seating	2 2	Softwood strips	
j	Base board	1	Exterior quality Ply or Oil temp'd hardboard	
k	Corner bracing	4	Exterior quality Plywood (triangles)	

Table 12.1 Component List for Solar Collector (the last column has been left blank for your own dimensions).

FIXINGS

Location	No. Reqd.	Item	Size mm	inches
Glazing and corner angles	26	Screws–round headed, aluminium	25	1
Base and corner braces	26	Screws, aluminium or stainless steel	25	1
Base seating, corner joints, glazing angle spacer and absorber locators	28	Nails–oval	30	1¼
Absorber supports	8	Screws	50	2
Corner joints	—	Wood glue		
Glass/Aluminium joint	1 tube	Silicone sealant (plus primer when recommended by manufacturer)		
All wood		Wood preservative		

Table 12.2 Fixings list for Solar Collector

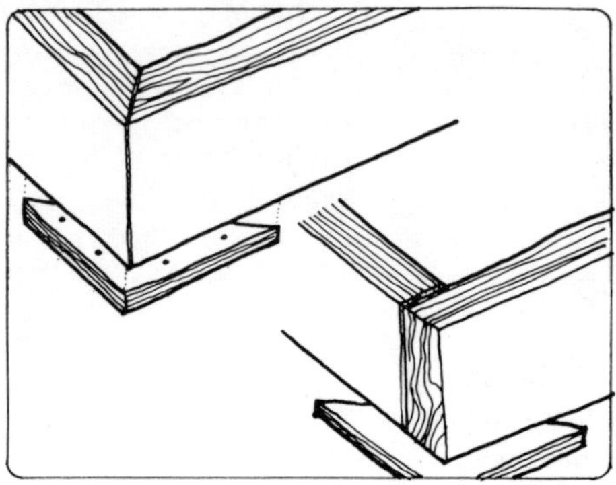

12.2 *Mitre or butt joints reinforced with plywood corner gussets*

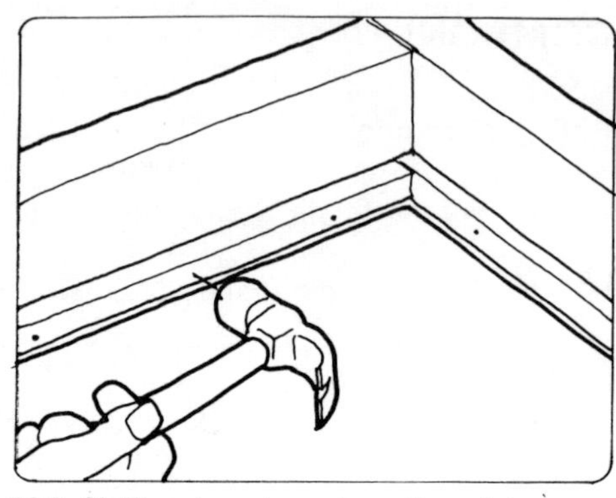

12.3 *Nailing baseboard seating strips*

ions of each component must be worked out. (These can then be listed against each component in the blank column in Table 12.1).

Construction is started by marking out the dimensions on the baseboard (¼″ exterior quality ply) using a straight-edge and square. Check that it is a true rectangle by ensuring that the diagonals are equal before sawing it out.

The side panels (normally of 2″ × 6″ planed redwood) are cut to fit round the outside of the baseboard. Mitred or butt joints are adequate, providing the faces are glued and the corners reinforced with plywood gussets screwed into the underside of the panels where they meet. (see 12.2)

The baseboard seating strips are now nailed to the sides so that the upper edge is at the correct height to receive the bottom of the absorber supports. The absorber supports span between the two long side panels, and are fixed by dowels and wood glue to the side panels. Always remember when using wood glue to wipe away any excess glue from the joint with a damp cloth.

The absorber supports should be situated so as to leave a gap of at least 3″ in the casing corners to allow for manipu-

12.4 *Absorber supports inside casing*

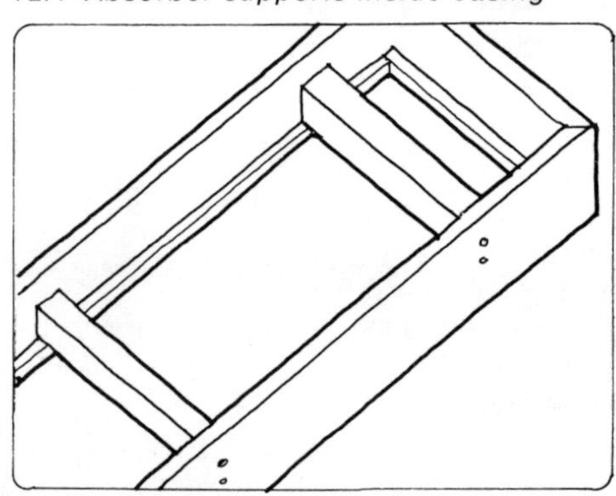

12.5 *After carefully marking absorber inlet and outlet positions, drill holes for connecting pipes*

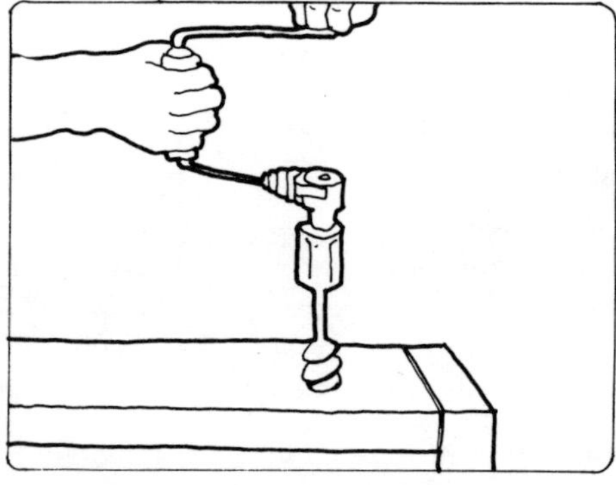

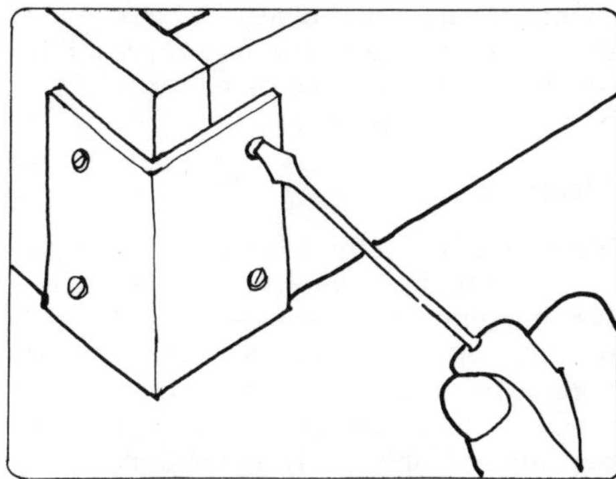

12.6 Screw on corner angles

12.7 Extruding a thin line of silicone sealant along the edge of the glass; mitre joints for the glazing angle and screwing to side panels

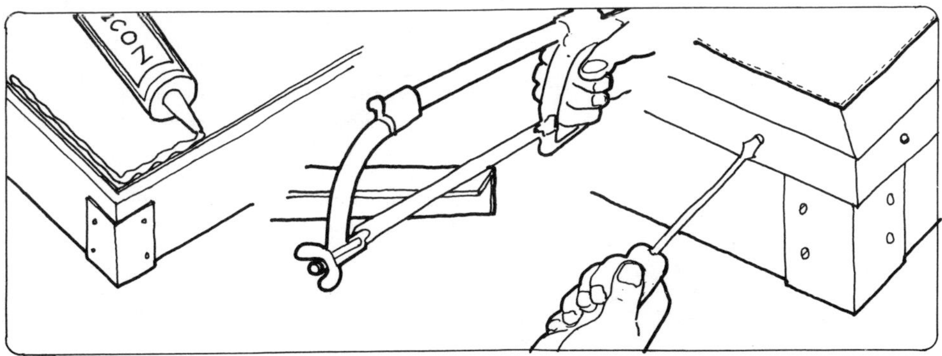

lation of the absorber connections. Also allow for the thickness of the absorber plus an air gap of at least ½", between the top of the support and the top of the casing.

The baseboard can now be screwed on if the assembly is turned upside down. If the base is going to be exposed to driving rain, it is important to bed the base in a sealing mastic in order to keep the insulation dry.

A 3" deep layer of mineral wool or glass fibre quilt can now be laid inside the box, turning it up at the edges to reduce heat losses through the sides. These are the materials usually used for roof insulation, and if you have ever used them you will remember that they can cause temporary irritation to the skin. Gloves and face mask should be worn when handling these materials.

You can now place the absorber plate inside the casing. Centre it upon its supports, and mark the inlet and outlet positions accurately on the side panels. The holes to accommodate these pipes can then be drilled.

With the inlet and outlet holes prepared, the aluminium corner angles can be screwed in place. These will protect the exposed end grain of the tubes and reinforce the framework. For added protection against the weather, they should be bedded in mastic. They should be located flush with the underside of the side panels, but with an allowance at the top for the glazing angle.

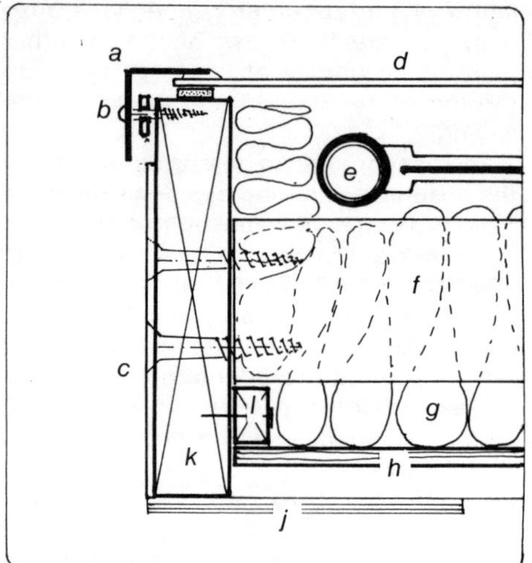

12.8 Cross-sectional view of solar collector
a) glazing angle
b) stainless steel screw with spacing washer
c) corner angle
d) glass
e) absorber
f) absorber support
g) insulation
h) base board
j) plywood corner bracing
k) side panel of casing
l) base seating

Once the absorber plate has been made and installed (see below for construction details) the glazing can be added and the installation completed.

Glazing

The cheapest method is traditional putty glazing. Under conditions of high temperature and direct sun however, this will need frequent replacement. A silicone sealant is more expensive but much longer lasting; however, subsequent removal of the glass may become difficult. The best alternative is the use of a neoprene based glazing tape, which is available with a self adhesive backing.

The glass can now be laid in place on top of the tape. Check that the casing is providing a level support all round the glass otherwise it may be cracked when the glazing angles are screwed down.

The right angle section aluminium which forms the glazing angle, should be cut to form a mitred joint at the corners. To do this accurately you need a 45° angle (or mitre box). The angle can be cut with a small hacksaw and the vertical side of the angle can be mitred with a file.

Screw holes should be drilled in the vertical side of the angle at the middle and near each end. (Make sure you do not drill through the horizontal surfaces!) Paraffin or turpentine can be used as a lubricant when drilling through aluminium.

Before laying the aluminium angle in place, it may be necessary to paint the inner surface of each piece with a special primer to ensure adhesion with the silicone sealant. Instructions on suitable primers should be given by manufacturers of the sealant.

With the glazing angle in place, press firmly downwards while attaching the angle to the side of the casing. Round headed screws and washers should be used, so that the angle is held away from the sides of the casing to form a drip edge.

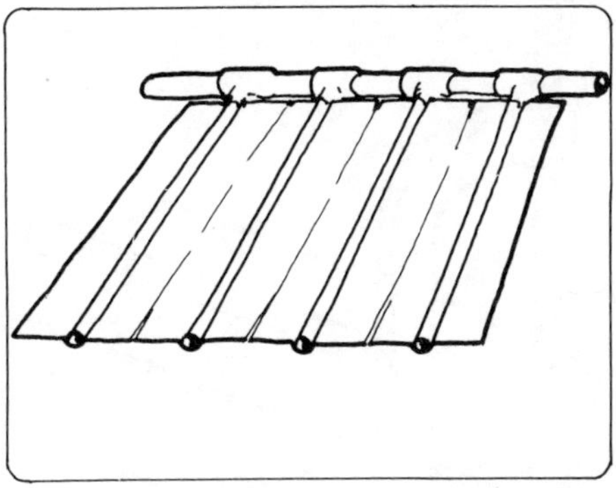

12.9 *Parallel grid absorber*

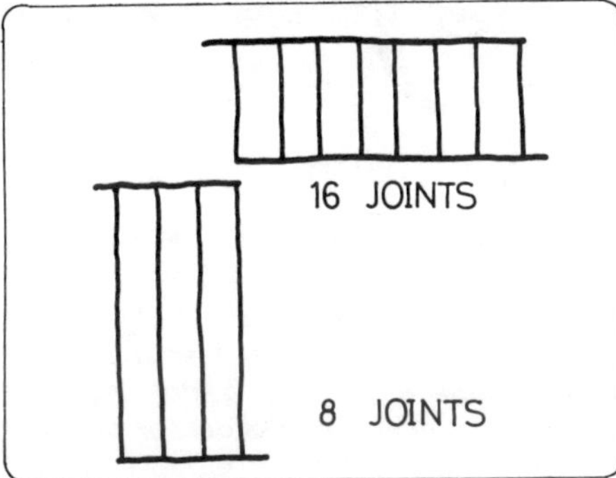

12.10 *The ratio of height to width determines the number of expensive tee fittings in a parallel grid absorber*

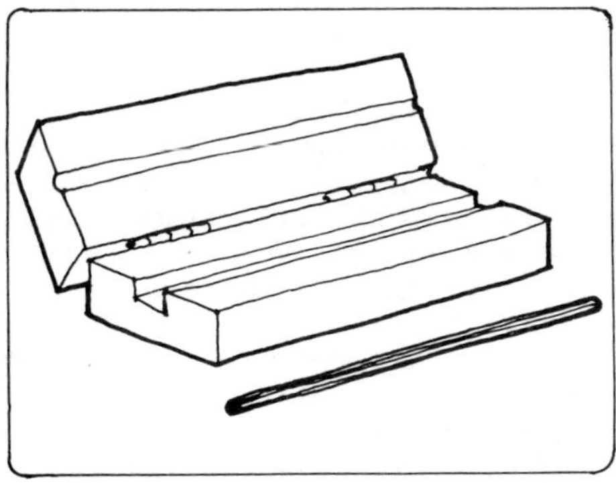

12.11 *Wood former for shaping copper sheet*

Construction of absorber plates

There are two types of tube and sheet absorber plates: parallel grid and serpentine configuration. Both types will operate with a pump; however if the system is to operate without a pump (gravity circulation) then the grid type is best. The serpentine type is inexpensive and quick to fabricate since it has few joints. The number of tee fittings required per unit area of absorber area in the grid type can be minimized by dimensioning it so that the distance between the two headers is as long as possible (see 12.10).

A parallel grid absorber plate is normally made from ½″ diameter vertical tubes and 1″ headers. Each vertical tube is bonded to an absorber sheet before assembly of the grid. The sheet (usually 6″ wide copper strip) is soldered to the vertical pipes.

Where circular section pipes are used, ribs are formed in the copper sheet using a timber former and ½″ diameter steel bar (12.11). The steel bar is placed on the centre of the copper sheet over the groove in the former, the top half of which is then closed over the bar and sheet and clamped, pressing the steel bar into the sheet and forming the U-section. The copper tube and ribbed sheet are then thoroughly

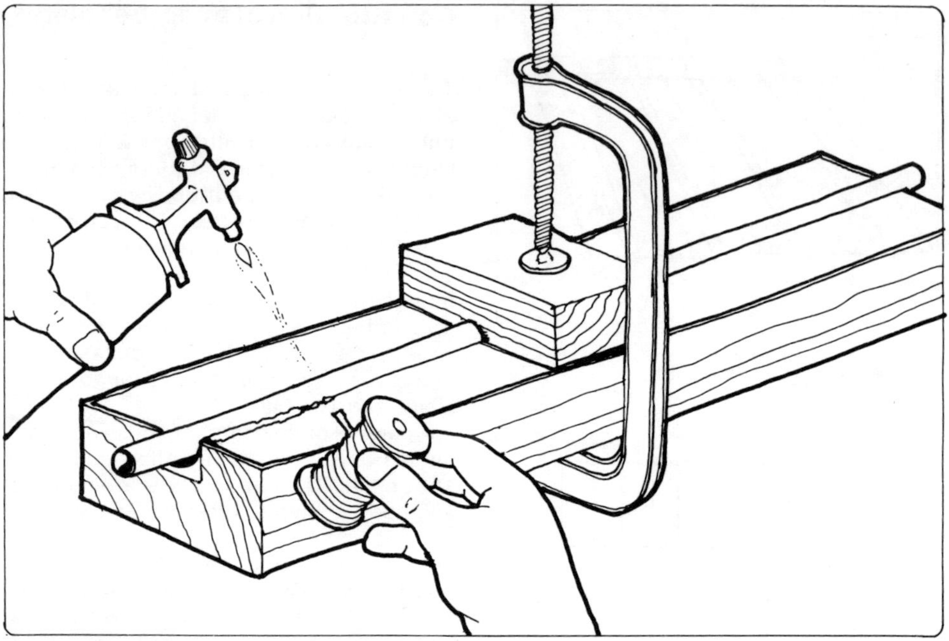

12.12 Grooved wood former for use in soldering tubes and sheets together. For good thermal contact it is essential that the tube and sheet are clamped tightly together during this process. Space G clamps on about 6" centre, solder between them then slide clamps along and solder the area which was inaccessible under the clamps.

12.13 Cross-sections of soldered round tube and D-section tube

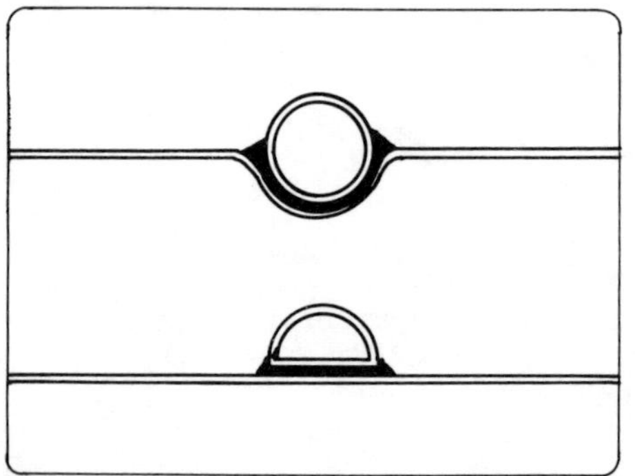

cleaned steel wool, and then placed in a second grooved wood former (see 12.12) to clamp tube and sheet tightly together during soldering. This ensures a good thermal connection between tube and sheet, and minimises the amount of solder needed.

The surfaces to be joined are first smeared with flux (see Chapter 8) and then heated using a propane torch. When the copper is hot enough, a strip of solder dipped in flux will melt into the groove. The joint should be left to cool before reclamping and moving on to the next section.

An alternative to forming ribs in the copper sheet is to use D-section copper tube, and solder straight onto the flat sheet (see 12.13). This is more straightforward, but the special fittings required may increase the cost.

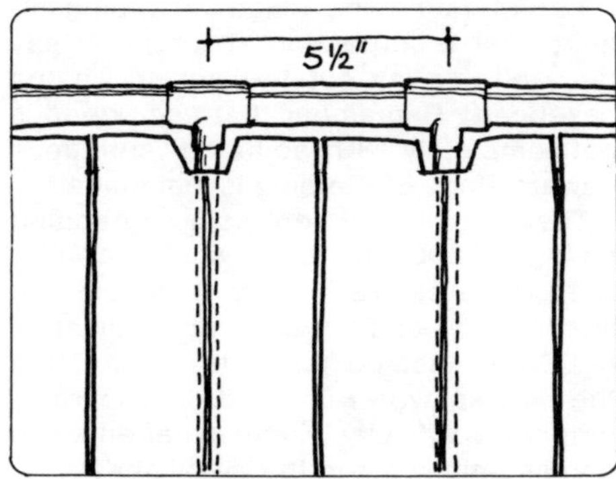

12.14 Parallel grid absorber showing the spacing of the T fittings

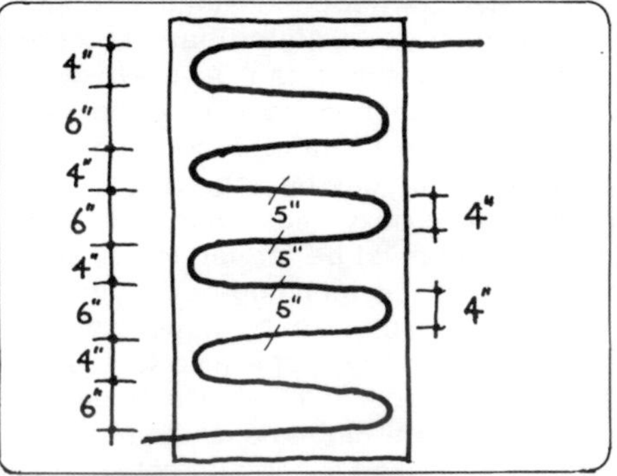

12.15 Serpentine absorber showing spacing arrangements

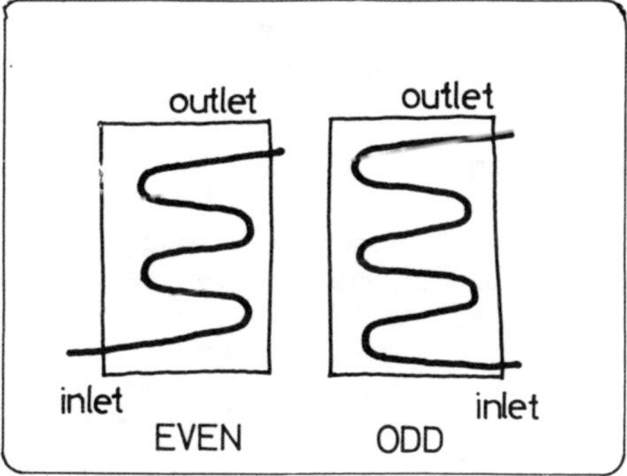

12.16 The number of bends in a serpentine absorber determine inlet and outlet positions

The headers are connected to the vertical tubes using reduced–T endfeed fittings (see Chapter 8). It is important to remember that solder used for the construction of the absorber plate should be 95:5 Tin/Antimony as this will tolerate high stagnation temperatures. Elsewhere in the system 50:50 Tin/Lead can be used.

Serpentine Tube Absorber

The construction of a serpentine tube absorber plate is simpler than the parallel grid, as it is basically one continuous length of copper pipe fixed to a single metal sheet or to a number of metal strips. Soft copper tube (Type L) should be used, as it is very difficult to bend rigid copper without heating it.

If you are building more than one collector, you need to think about how you are linking up each of the collectors as this will influence the number of bends in the tube: an even number of bends will result in the inlet and outlet being on the opposite sides of the casing, an odd number will result in both being on the same side (see fig 12.16 and Chapter 10).

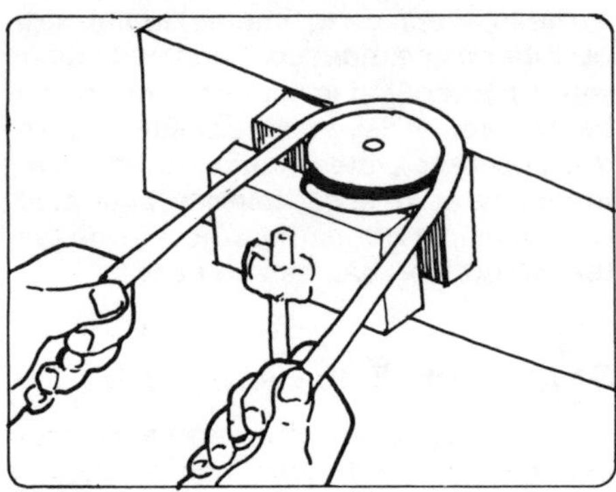

12.17 Bending a pipe with a pulley fixed in a vice

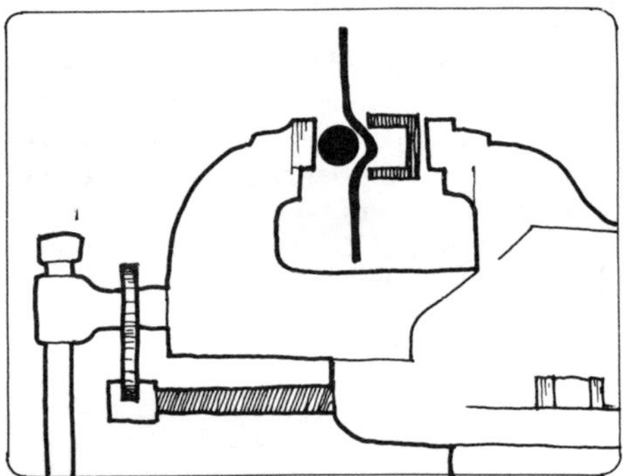

12.18 Bending aluminium sheet around copper tube prior to glueing

12.19 Aluminium strips bonded to serpentine copper tube

To determine the length of tubing required for a serpentine absorber, it may be useful to lay out the design on the baseboard. Remember that each bend is not completely 180°, so that a continuous upward slope of the pipe is maintained.

The tube can be bent using a bending spring and a suitable former such as a pulley fixed in a vice (12.17). The minimum diameter of the bend should be approximately 4″. Care must be taken to ensure that the bent pipework remains flat in one plane, otherwise difficulty will be experienced in bonding pipework to the metal sheet.

The serpentine pipe is bonded either to a single large metal sheet or to a number of metal strips. For a do-it-yourself solar collector it is easier to use several strips, as this allows greater dimensional tolerance in construction. As stated earlier, copper or aluminium sheet can be used. When aluminium sheet is used (second hand printing plates will do) the bond between sheet and tube can be achieved by using a two-part Epoxy glue. But first semi-circular ribs must be formed in the sheet either using the technique described for the grid absorber or by squeezing the sheet between a U–profile mild steel extrusion and a steel bar using a large vice (see 12.18). The ribbed aluminium sheets can then be fixed to the straight sections of the serpentine pipe using the Epoxy glue. The two components of the glue are thoroughly mixed together, spread over each of the joining surfaces allowed to partially cure for a few minutes and then firmly clamped together.

Painting

One side of the absorber plate must now be painted to increase its absorbtivity. First clean the surface thoroughly to remove dirt and grease. Next a self etching primer should be applied to ensure a lasting contact between the metal and the paint. When the primer is dry coat the plate with a thin layer of matt black paint. There are special coating materials produced for this purpose, such as 3M's Nextel. You could also buy a

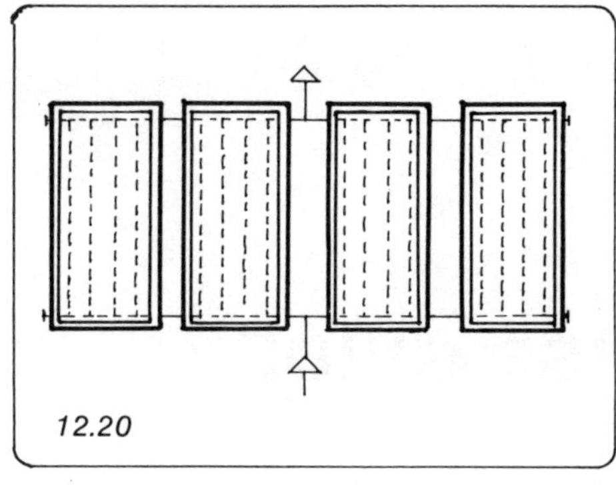
12.20

selective coating which comes in the form of adhesive foil strips. But these require a perfectly flat mounting surface and it is not easy to keep the sheet of flat metal free from bumps and wrinkles during the process of forming channels and bonding pipes.

Manufactured Components

There are now many manufactured items available which can make collector construction easier. One of the best guides to these materials is a catalogue produced by SOLAR USAGE NOW—a mail order company whose address is given in Appendix 3.

Of particular interest are the aluminium profiles for casing construction. The catalogue also features absorber coating, thermal cement, D-profile copper tube, and the synthetic rubber EPDM Solaroll Absorbers described in Chapter 4. I have been amazed to discover that the cost of buying these absorbers would sometimes have been cheaper than constructing my own from copper.

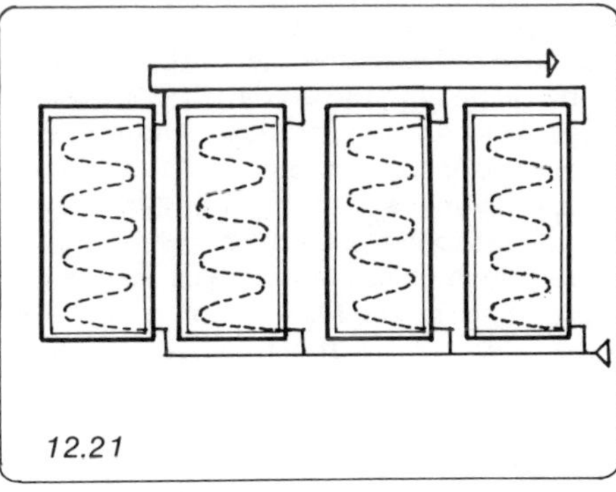
12.21

Mounting the Absorbers

The serpentine absorber should be mounted so that the tubes run from side to side, i.e. nearly horizontal, so that it is possible to drain the collectors without having water trapped in the zig-zag coils. Grid absorbers, however, should be mounted with their 15mm tubes running from top to bottom to facilitate gravity circulation through them. For detailed notes on the effects of different connection patterns, see Chapter 10.

The number of people in a household will clearly influence the number of solar collectors required to meet the demand for hot water. Also, the more collectors required, of course the more the system will cost. Up to a point however, the increase in energy collection that can be yielded by additional collectors will justify the extra expenditure. This is particularly so when using do-it-yourself panels which are comparatively cheap. This and other economic factors will be discussed in the next chapter which looks at the costs involved and compares these with benefits returned by solar systems.

13

A DOLLAR-WISE GUIDE

The installation of a solar heating system has many advantages some of which can be measured in financial terms, and others which cannot. Houseowners who install a system will have reduced fuel bills, move a step closer to self-sufficiency, increase the value of their house and set a valuable example to their community. The community itself will benefit because each solar system installed means that less fossil fuel will be burned and the air will be cleaner and the environment more healthy. The whole country benefits as each new solar system represents a saving in fuel imports and hence alleviates economic problems caused by trade deficits. Examining effects on an even wider level, the whole world can be seen to share the rewards. The development of solar technology will help conserve the precious fuel reserves still remaining, thereby extending their benefits to future generations and easing the tensions which will undoubtedly arise between nations if there is a sudden and drastic cut in the accustomed levels of energy supply.

The advantages which are perhaps the most important are also the most difficult to quanitify. How do you fix a price tag on clean air? How many B.t.u.'s of solar energy are required to prevent the outbreak of a resource war between energy hungry countries? We live in a society which increasingly tends to base its decisions upon quantified data and arguments based on the idea of quality tend to be ignored.

Between 1972 and 1981 the price of crude oil has increased by 1800%. Such a massive and rapid increase has made it much easier to demonstrate the financial competitiveness of solar heating. But, more significant I believe, has been the change in attitudes which has taken place across the same time span. Millions of Americans have experienced the frustration, inconvenience and deep anxiety caused by fuel shortages and nuclear power plant accidents. There is now a strong demand for moving the economy away from reliance on instable and unsafe energy sources.

Our society is coming to appreciate the "hidden" benefits of using a constantly renewed and reliable source of energy, at the same time as economists are beginning to report that solar water heating can be cost competitive with conventional fuels almost anywhere in the country.

The remainder of this chapter is aimed at

helping you to evaluate the value of an expenditure on solar water heating, in the same way that you might evaluate an investment. It leaves you to evaluate the "hidden" benefits for yourself.

The Solar Investment

In financial terms, a solar system is a capital investment which reduces recurrent expenditure (fuel bills). The return on the investment is the saving on the fuel bill.

Tax Credits

An important advantage of a solar investment is that the returns will be tax free as they come in the form of savings rather than actual payments. Current legislation grants further benefits to solar investments through the tax system. The Windfall Profits Tax Legislation provides for tax credits of 40% on residential solar heating systems up to a maximum credit of $4000. This has the effect of reducing the cost without any decrease in the financial returns. In addition, at least 41 states have enacted their own tax break programs, some of which can cut away further at the cost of the system (see Chapter 14).

Financial Analysis

An accurate evaluation of the future savings from a solar system is not easy because of the difficulties inherent in predicting the future. A great variety of financial jargon will be encountered in any of the detailed analyses which attempt this, but it is worth remembering that no final and definite answers can result. What we get is a mixture of calculated guesswork plus a touch of crystal ball gazing which will indicate some possible results.

Our first analysis will be a simple payback-period calculation, which is crude but quick. It will be followed by a more detailed study of year–by–year cash flows.

Both methods initially require an estimation of fuel bill savings, and system cost.

Step 1 Fuel Bill Savings

The amount by which your water heating bill will be reduced depends on many factors; the climate, the area of solar panels you install, the efficiency of the panels, the amount of hot water your household requires, the temperature at which it is required and the temperature at which cold water is delivered are the major influences.

If you want to spend a lot of time and make as accurate an estimate as is possible without turning to a computer, you could use the technique presented in "Solar Heating Design–f chart method by Duffie and Beckman.

For a much quicker estimate you can refer to table 13.1 which lists the units of fuel you might expect to have replaced by various sizes of solar system. First check through the locations for the city nearest to your site. Next select the column indicating hot water requirements. (The average American uses about 20 gallons per day.) The energy saved by the solar system is specified both in electric units (kWh) and natural gas units (cubic feet). The figures have been calculated for a solar system with the following characteristics:

Collector type	:	single glazed, black-painted absorber.
Collector orientation	:	within 15° of south
Collector tilt	:	equal to latitude of the site
Cold water temp.	:	55°F
Hot water temp.	:	140°F
Efficiency of boiler	:	gas units replaced are calculated on the basis of a 70% boiler efficiency. Electric boilers are assumed to be 100% efficient.

The performance of your system could be better, or worse, than the table indicates if it has very different characteristics.

Table 13.1

Location	Collector Area (sq.ft)	Annual Total of Energy Units Saved by Solar System: Hot Water Req. 60 galls Electricity (kWh)	Hot Water Req. 60 galls Nat.Gas (ccf)	Hot Water Req. 80 galls Electricity (kWh)	Hot Water Req. 80 galls Nat.Gas (ccf)	Hot Water Req. 120 galls Electricity (kWh)	Hot Water Req. 120 galls Nat.Gas (ccf)
ALABAMA Birmingham	30	1991	97	2052	100	2172	106
	60	3304	161	3620	176	3983	194
	120	4254	207	5370	262	6608	321
ARIZONA Phoenix	30	2761	134	2957	144	3078	150
	60	4300	209	4948	241	5522	268
ARKANSAS Fort Smith	30	1946	95	2052	100	2082	101
	60	3259	158	3560	173	3892	189
	120			5310	259	6517	317
CALIFORNIA Fresno	30	2263	110	2414	118	2535	123
	60	3666	178	4103	200	4526	220
	120			5491	268	7332	356
CALIFORNIA Inyokern	30	3168	154	3379	165	3621	176
	60	4435	216	5491	268	6336	308
CALIFORNIA Sacramento	30	2218	108	2353	115	2444	119
	60	3576	174	3982	194	4435	216
	120	4209	205	5370	262	7151	348
CALIFORNIA San Diego	30	2082	98	2172	106	2263	110
	60	3485	169	3801	185	4164	202
	120			5732	279	6970	339
CALIFORNIA San Francisco	30	2037	99	2112	103	2263	110
	60	3349	163	3681	179	4073	198
	120			5491	268	6698	326
CALIFORNIA Santa Maria	30	2489	121	2655	129	2806	136
	60	3983	194	4465	218	4979	242
	120			5913	288	7966	387
COLORADO Denver	30	2444	119	2534	123	2716	132
	60	3892	189	4344	211	4888	238
	120			5913	288	7784	378
CONNECTI-CUT Hartford	60	2942	143	3198	156	3440	167
	120	4073	198	5008	244	5884	286
DIST of Columbia Washington	60	2896	140	3137	152	3439	167
	120	4028	195	4887	238	5793	281
FLORIDA Jacksonville	30	2036	99	2111	102	2263	110
	60	3439	167	3741	182	4073	198
	120	4435	215	5611	273	6879	334
FLORIDA Miami	30	2308	112	2413	117	2534	123
	60	3847	187	4223	205	4616	224
	120					7694	374
FLORIDA Pensacola	30	2081	101	2172	105	2263	110
	60	3485	169	3801	185	4163	202
	120	4390	213	5611	273	6970	338
FLORIDA Tallahassee	30	2172	105	2292	111	2353	114
	60	3620	176	3982	194	4344	211
	120	4480	217	5852	285	7241	352
FLORIDA Tampa	30	2353	114	2473	120	2534	123
	60	3847	187	4223	205	4707	228
	120			5973	291	7694	374
GEORGIA Atlanta	30	1991	96	2051	91	2172	105
	60	3303	160	3620	176	3982	193
	120			5490	267	6607	321

Location	Solar Collector Area (sq.ft)	Annual Total Energy Units Saved By Solar System					
		Hot Water Daily Req. 60 galls		Hot Water Daily Req. 80 galls		Hot Water Daily Req. 120 galls	
		Electricity (kWh)	Nat.Gas (ccf)	Electricity (kWh)	Nat.Gas (ccf)	Electricity (kWh)	Nat.Gas (ccf)
IDAHO	30	2036	99	2172	105	2263	110
Boise	60	3349	162	3741	182	4073	198
	120			5249	255	6698	325
IDAHO	30	2172	105	2292	111	2444	118
Pocatello	60	3530	171	3922	191	4344	211
	120			5430	264	7060	343
ILLINOIS	60	2941	143	3198	155	7513	365
Chicago	120	4028	195	4947	241	5883	286
ILLINOIS	60	3032	147	3318	161	3620	176
Peoria	120	4028	195	5008	244	6064	294
INDIANA	60	2851	138	3077	149	3349	162
Indianapolis	120	3892	189	4766	232	5702	277
INDIANA	60	2987	145	3258	158	3530	171
Fort Wayne	120	3982	193	4947	241	5974	290
IOWA	60	3077	149	3379	164	3711	180
Des Moines	120	4073	198	5068	246	6155	299
KANSAS	60	3847	187	4284	208	4797	233
Dodge City	120					7694	374
KANSAS	60	3485	169	3861	188	4254	206
Wichita	120	4344	211	5551	270	6970	338
KENTUCKY	60	3439	167	3801	185	4163	202
Lexington	120	4209	204	5370	261	6879	334
LOUISIANA	30	2081	101	2172	105	2263	110
Lake Charles	60	3485	169	3801	185	4163	202
	120	4390	213	5611	273	6970	338
MAINE	60	2715	132	2956	144	3258	158
Caribou	120	3847	187	4646	226	5431	264
MAINE	60	3032	147	3318	161	3620	176
Portland	120	4118	200	5068	246	6064	294
MASSACHU-SSETTS	60	2625	127	2835	138	2987	145
Boston	120	3801	184	4465	217	5250	255
MICHIGAN	60	2760	134	3017	147	3258	158
Detroit	120	3801	184	4646	226	5521	268
MINNESOTA	30	1674	81	1749	85		
Minneapolis	60	2806	136	3077	149	3349	162
	120	3937	191	4766	232	5612	272
MISSISSIPPI	30	1991	96	2051	99	2172	105
Jackson	60	3303	160	3620	176	3982	193
	120			5430	264	6607	321
MISSOURI	30	1946	94	2051	99	2081	101
Kansas City	60	3258	158	3560	173	3892	189
	120	4209	204	5249	255	6517	316
MONTANA	30	2036	99	2111	102	2263	110
Billings	60	3349	162	3680	179	4073	198
	120	4254	206	5309	258	6698	325
NEBRASKA	30	1946	94	1991	97		
Lincoln	60	3213	156	3560	173	3892	189
	120	4254	206	5309	258	6426	312
NEVADA	30	2760	134	2956	144	3077	149
Las Vegas	60	4299	209	4887	238	5521	268
	120					8599	418
NEVADA	30	2534	123	2715	132	2896	140
Reno	60	4028	194	4525	220	5069	246
	120			5852	285	8056	391

Location	Solar Collector Area (sq.ft)	Annual Total Energy Units Saved By Solar System Hot Water Daily Req. 60 galls		Hot Water Daily Req. 80 galls		Hot Water Daily Req. 120 galls	
		Electricity (kWh)	Nat.Gas (ccf)	Electricity (kWh)	Nat.Gas (ccf)	Electricity (kWh)	Nat.Gas (ccf)
NEW JERSEY	60	3168	154	3439	167	3801	184
Atlantic City	120	4209	204	5249	255	6336	308
NEW MEXICO	30	2760	134	2956	144	3077	149
Albuquerque	60	4299	209	4887	238	5521	268
	120			5973	291	8599	418
NEW YORK	60	2806	136	3077	149	3349	162
Albany	120	3892	189	4706	229	5612	272
N.CAROLINA	30	1946	94	1991	97	2081	101
Greensboro	60	3258	158	3560	173	3892	189
	120			5370	261	6517	316
N.CAROLINA	30	2308	112	2413	117	2534	123
Cap Matteras	60	3756	182	4163	202	4616	224
	120			5792	282	7513	365
N.DAKOTA	30	1719	83	1749	85	1810	88
Fargo	60	2896	140	3137	152	3439	167
	120	3937	191	4827	235	5793	281
OHIO	60	2670	129	2896	141	3168	154
Cleveland	120	3666	178	4525	220	5340	259
OKLAHOMA	30	2263	110	2413	117	2534	123
Oklahoma	60	3711	180	4103	199	4526	220
City	120			5792	282	7422	360
OREGON	60	2353	113	2534	123	2715	132
Portland	120	3439	167	4103	199	4707	228
OREGON	30	1900	92	1991	97	2081	101
Medford.	60	3122	151	3439	167	3801	184
	120	3801	184	4827	235	6245	303
PENNSYL-VANIA	60	3032	147	3318	161	3620	176
	120	4118	200	5128	249	6064	294
S.CAROLINA	30	2081	101	2172	105	2263	110
Charleston	60	3439	167	3741	182	4163	202
	120			5611	273	6879	334
S.DAKOTA	30	2172	105	2292	111	2444	118
Rapid City	60	3575	173	3982	194	4344	211
	120	4390	213	5671	276	7151	347
TENNESSEE	30	1900	92	1991	97	2081	101
Chattanooga	60	3213	156	3499	170	3801	184
	120	4209	204	5249	255	6426	312
TEXAS	30	2489	121	2654	129	2806	136
Amarillo	60	3982	193	4465	217	4978	242
	120			5913	288	7965	387
TEXAS	30	2806	136	2956	144	3168	154
El Paso	60	4344	211	4947	241	5612	272
TEXAS	30	2081	101	2172	105	2263	110
Houston.	60	3485	169	3801	185	4163	202
	120			5551	270	6970	338
UTAH	30	1991	96	2111	102	2172	105
Salt Lake City	60	3303	160	3680	179	3982	193
	120	4163	202	5249	255	6607	321
VIRGINIA	60	3168	154	3439	167	3711	180
Richmond	120	4209	204	5249	255	6336	308
WISCONSIN	30	1719	83	1749	85	1810	88
Milwaukee	60	2896	140	3137	152	3439	167
	120	3937	191	4827	235	5793	281

Location	Solar Collector Area (sq.ft)	Hot Water Daily Req. 60 galls Electricity (kWh)	Hot Water Daily Req. 60 galls Nat.Gas (ccf)	Hot Water Daily Req. 80 galls Electricity (kWh)	Hot Water Daily Req. 80 galls Nat.Gas (ccf)	Hot Water Daily Req. 120 galls Electricity (kWh)	Hot Water Daily Req. 120 galls Nat.Gas (ccf)
		Annual Total Energy Units Saved By Solar System					
WASHING-	30	1403	68	1448	70		
TON	60	2444	118	2594	126	2806	136
Seattle	120	3394	165	4103	199	4888	237
WASHING-	30	1855	90	1930	94		
TON	60	3032	147	3379	164	3711	180
Spokane	120	3847	187	4887	238	6064	294
WYOMING	30	2172	105	2292	111	2353	114
Laramie	60	3530	171	3922	191	4344	211
	120			5671	276	7060	343
W.VIRGINIA	60	2760	134	3017	147	3258	158
Parkersburg	120	3847	187	4706	229	5521	268
CANADA							
BRITISH	60	1991	96	2172	105	2263	110
COLOMBIA	120	2987	145	3499	170	3982	193
Vancouver							
MANITOBA	60	2987	145	3258	158	3620	176
Winnipeg	120	4028	195	4947	241	5974	290
NEW	60	2353	114	2534	123	2715	132
BRUNSWICK	120	3485	169	4042	196	4707	228
QUEBEC	60	2444	118	2594	126	2806	136
Montreal	120	3485	169	4163	202	4888	237

Having consulted table 13.1, you now have to check with your local utility company to find the current price of fuel. Fill in the values on worksheet (A) to obtain an estimate of the fuel bill saving due to the solar system.

Worksheet (A)

Units of Fuel Saved (table 13.1)		3198
Price/Unit of Fuel (from utility) $	X	0·07
Annual Fuel Bill Saving $	=	223·86

There is a collection of blank worksheets at the end of this chapter to enable you to figure out your own system. The worksheets shown here in the text have been filled out, by way of example, by a 4-person family living in Connecticut: Based on averages, we assume they use 80 gallons of hot water per day and they have installed 60 square feet of solar collectors. As they are using electricity as a back-up fuel, we have recorded on worksheet (A) the amount of fuel saved in kilowatt-hours. If their utility company charged them 7¢ per kWh, their solar system could save them $223 a year.

Step 2. System Cost

The basic cost can be found by asking for quotations from local installers, or, if you are doing the work yourself, by making out a shopping list with reference to the previous chapters.

A typical 60 square foot system would have cost about $3000 in 1980. This amount can be reduced to $1800 if you claim the 40% federal tax credit, and perhaps even further if your state provides cumulative state tax credits for solar users.

Step 3. Payback Period

The payback period is quite simply the number of years your solar system has to operate before the fuel savings it provides equal the cost of the system. Thereafter any solar savings can be considered as profit. Dividing the system cost by the annual fuel bill saving gives a quick indication of the payback period.

Worksheet (B)

System Cost (Step 2) $		1800·00
Annual Fuel Bill Saving (Step 1) $	÷	223·86
Payback period yrs.	=	8

If the payback period calculated in this first stage is short, you may decide not to bother with further reckoning. If the period is long however, there is a danger that large errors will have built up due to our ignoring the effects of full price increases, interest rates and inflation.

Step 4. Fuel Price Increases

Now we come to the big guessing game. How much will fuel prices rise, over and above general inflationary increases? In the example below, we will use a rate 7½% each year. Try calling up your local utility and check on recent trends and their predictions ... then get out your own crystal ball.

Table 13.2 shows how the value of $1.00 increases at different rates. You can use these values as "interest factors" in worksheet (C) to find your future fuel bill savings at your chosen rate of increase.

Worksheet (C)

Year Number	Initial Fuel Bill Saving $ (from worksheet (A))		Interest Factor @ 7½ % (from table 13.2)		Future Fuel Bill Savings $
1	223·86	X	1·075	=	241
2		X	1·156	=	259
3	"	X	1·242	=	278
4		X	1·335	=	299
5		X	1·436	=	321
6		X	1·543	=	345
7		X	1·659	=	371
8	"	X	1·783	=	399
9		X	1·917	=	429
10		X	2·061	=	461
11		X	2·216	=	496
12		X	2·382	=	533
13	"	X	2·560	=	573
14		X	2·752	=	616
15		X	2·959	=	662

Number of Years	Rates of Interest 5%	7½%	10%	15%
1	1.050	1.075	1.100	1.15
2	1.103	1.156	1.210	1.323
3	1.158	1.242	1.331	1.521
4	1.216	1.335	1.464	1.749
5	1.276	1.436	1.611	2.011
6	1.340	1.543	1.772	2.313
7	1.407	1.659	1.949	2.660
8	1.477	1.783	2.144	3.059
9	1.551	1.917	2.358	3.518
10	1.629	2.061	2.594	4.046
11	1.710	2.216	2.853	4.652
12	1.796	2.382	2.138	5.350
13	1.886	2.560	3.452	6.153
14	1.980	2.752	3.798	7.076
15	2.079	2.959	4.177	8.137

Table 13.2 "Interest Factors"

This table shows the increase in value of $1.00 at different rates of interest.
Note: If the table does not show the interest rates that you require, you can calculate your own "interest factors" from the following formula:

Interest Factor = Initial sum ($1) Plus Interest after (y) years @ (r) rate of interest = $\$1.00(1+r)^y$

Step 5. Operating Costs

If you are planning to have a pumped solar system, you will have to pay running costs. In 1980, a typical annual cost for electricity to run a pump would be $10.00. This amount would increase each year at the same rate as the electricity price rises. Reference to table 13.2 will give appropriate interest factors to enable you to complete worksheet (D).

Worksheet (D)

Year Number	Initial Running Cost $		Interest Factor @7½%		Annual Running Cost $
1	10·00	X	1·075	=	11
2		X	1·156	=	12
3	"	X	1·242	=	12
4		X	1·335	=	13
5		X	1·436	=	14
6		X	1·153	=	15
7		X	1·659	=	17
8	"	X	1·783	=	18
9		X	1·917	=	19
10		X	2·061	=	21
11		X	2·216	=	22
12		X	2·382	=	24
13	"	X	2·560	=	26
14		X	2·752	=	28
15		X	2·959	=	30

Step 6. Maintenance and Total Costs

Maintenance charges should not be necessary every year. But it is advisable to make allowance for the occasional expenditure. In our example we will allow for the renewal of the heat transfer fluid every five years, and the replacement of the pump, after ten years. These costs are shown on worksheet (E).

Worksheet (E)

Year Number	Initial Cost Minus Tax Credit $		(from worksheet (D)) Running Cost $		Maintainance $		Total Costs $
1	1800	+	11	+	0	=	1811
2		+	12	+	0	=	12
3		+	12	+	0	=	12
4		+	13	+	0	=	13
5		+	14	+	40	=	54
6		+	15	+	0	=	15
7		+	17	+	0	=	17
8		+	18	+	0	=	18
9		+	19	+	0	=	19
10		+	21	+	60	=	81
11		+	22	+	80	=	102
12		+	24	+	0	=	24
13		+	26	+	0	=	26
14		+	28	+	0	=	28
15		+	30	+	0	=	30

Step 7. Present Value of Cash Flows

Just as you can say that $1.00 today is worth $1.10 in a years time if it is invested at 10% interest rate, so you can also say that a saving of $1.10 next year, is worth only $1.00 in present-day value. The rate at which we reduce future sums is called the test discount rate. Choosing a test discount rate is another case for the crystal ball. It should be a rate which you can reasonably expect your money could earn if you invested or banked it instead of buying a solar system. Remember to deduct your estimate for general inflation from this expected interest rate. For example, if you think your money could earn 11% interest, and that general inflation will average out at 6%, then you would choose a test discount rate of 5%.

Worksheet (F)

Year Number	Future Fuel Bill Savings $		Total Costs $		Solar Savings $		Discount Divisor at 5 %		Present Value of Solar Saving $	Cumulative Total $
1	241	–	1811	=	-1570	÷	1·050	=	-1495	-1495
2	259	–	12	=	247	÷	1·103	=	224	-1271
3	278	–	12	=	266	÷	1·158	=	230	-1041
4	299	–	13	=	286	÷	1·216	=	235	- 806
5	321	–	54	=	267	÷	1·276	=	209	- 597
6	345	–	15	=	330	÷	1·340	=	246	- 351
7	371	–	17	=	354	÷	1·407	=	252	- 99
8	399	–	18	=	381	÷	1·477	=	246	147
9	429	–	19	=	410	÷	1·551	=	264	411
10	461	–	81	=	380	÷	1·629	=	233	644
11	496	–	102	=	394	÷	1·710	=	230	874
12	533	–	24	=	509	÷	1·796	=	283	1157
13	573	–	26	=	547	÷	1·886	=	290	1447
14	616	–	28	=	588	÷	1·980	=	297	1744
15	662	–	30	=	632	÷	2·079	=	304	2048

Once a test discount rate has been selected, a DISCOUNT DIVISOR can be read off table 13.2 in the same way that interest factors were obtained. When the annual solar saving is divided by the discount divisor, we obtain its present value. The present values of the solar savings can be added to gain a cumulative total of solar savings at present value. This gives us a better picture of cash flows and indicates the system's payback period more accurately.

In the example shown the solar system breaks even in its eighth year–just a few months earlier than our sample payback calculation indicated. The second method also provides more about the subsequent "profits" that might accrue from the solar investment.

Remember that the calculation has to be repeated now for your local conditions and incorporating your financial hunches rather than mine. So, start filling in those blank worksheets.

Worksheet (A)

Units of Fuel Saved (table 13.1)		
Price/Unit of Fuel (from utility) $	X	
Annual Fuel Bill Saving $	=	

Worksheet (B)

System Cost (Step 2) $		
Annual Fuel Bill Saving (Step 1) $	÷	
Payback period yrs.	=	

Worksheet (C)

Year Number	Initial Fuel Bill Saving $ (from worksheet (A))		Interest Factor @ % (from table 13.2)		Future Fuel Bill Savings $
1		X		=	
2		X		=	
3		X		=	
4		X		=	
5		X		=	
6		X		=	
7		X		=	
8		X		=	
9		X		=	
10		X		=	
11		X		=	
12		X		=	
13		X		=	
14		X		=	
15		X		=	

Worksheet (D)

Year Number	Initial Running Cost $		Interest Factor @ %		Annual Running Cost $
1		X		=	
2		X		=	
3		X		=	
4		X		=	
5		X		=	
6		X		=	
7		X		=	
8		X		=	
9		X		=	
10		X		=	
11		X		=	
12		X		=	
13		X		=	
14		X		=	
15		X		=	

Worksheet (E)

Year Number	Initial Cost Minus Tax Credit $		(from worksheet (D)) Running Cost $		Maintainance $		Total Costs $
1		+		+		=	
2		+		+		=	
3		+		+		=	
4		+		+		=	
5		+		+		=	
6		+		+		=	
7		+		+		=	
8		+		+		=	
9		+		+		=	
10		+		+		=	
11		+		+		=	
12		+		+		=	
13		+		+		=	
14		+		+		=	
15		+		+		=	

Worksheet (F)

Year Number	Future Fuel Bill Savings $		Total Costs $		Solar Savings $		Discount Divisor at %		Present Value of Solar Saving $	Cumulative Total $
1		–		=		÷		=		
2		–		=		÷		=		
3		–		=		÷		=		
4		–		=		÷		=		
5		–		=		÷		=		
6		–		=		÷		=		
7		–		=		÷		=		
8		–		=		÷		=		
9		–		=		÷		=		
10		–		=		÷		=		
11		–		=		÷		=		
12		–		=		÷		=		
13		–		=		÷		=		
14		–		=		÷		=		
15		–		=		÷		=		

14

TAX CREDITS & SUBSIDIZED LOANS

Public interest and support for solar energy has led to a flurry of activity among our law-makers. A great stack of new legislation has come into being to promote solar investment, subsidize loans, ease zoning codes and protect our access to sunlight.

These new laws make up a very generous financial incentive for going solar. However we are already beginning to build up a brand new maze of rules and regulations. Claiming one benefit may rule out the possibility of claiming another. Before making a final choice about the solar system you buy, you ought to investigate how the various incentives apply to you. There are usually standards set out by the official body concerned relating to the equipment installed. If a particular subsidy is worth a lot to you, you will want to ensure that your equipment has the required mark of approval.

Federal Tax Credit

Starting with federal laws, the most important is the Crude Oil Windfall Profit Tax Act of 1979. This provides for a non-refundable tax credit of 40% of the cost of buying and installing renewable energy systems in both new and existing buildings. The maximum credit you can claim is $4000. To qualify for this credit your solar investment will have to be made before 31 December 1985. If the claim you make is actually higher than the tax you owe, credits can be extended over future years through until 1987. Details and appropriate forms can be obtained from your local Internal Revenue Service office.

The "Provisions to Prevent Double Benefit" which form section 203 of the Windfall Profit Act may reduce or eliminate the 40% tax credit if you have made use of low interest loans subsidized by federal, state or local government as part of an energy-saving program. In most cases, the federal tax credit is worth more than subsidized loans. However you might wish to check your own particular case.

Solar Bank

Another important federal initiative has been the setting up of the Solar Energy and Energy Conservation Bank, as a corporation within the U.S. Department of Housing and Urban Development. The intention here was to make it easier for people to obtain the necessary capital to make an investment in a solar system. This is to be done by effectively lowering the interest rates on money borrowed for solar projects. The Solar Bank does not itself make loans. You have to borrow from the usual sources, banks or financial institutions. The Solar Bank will subsidize your interest repayments. The exact details have not been disclosed at the time of writing, but up to $5,000 worth of subsidy may be available through this source. As mentioned above, it is not possible to claim both a tax credit, and a subsidized loan for the same project. For water heating systems, the tax credit will probably be the most valuable option for the majority of householders.

State Tax Credits

Many states now offer personal income tax credits to individuals installing solar systems. These vary greatly in amount, and in their relation to federal tax credits. In New Mexico, for example, you can claim a credit equal to 25% of the cost of a solar system, in addition to any federal tax credit. Whereas in California, the federal tax credit is deducted from the state credit which is worth 55% of the system cost.

The following states have income tax credit programmes covering residential solar water heating installations.

Alaska
Arkansas
California
Colorado
Delaware
Hawaii
Idaho
Kansas
Michigan
Montana
New Mexico
North Carolina
North Dakota
Oklahoma
Oregon
Vermont
Wisconsin

Contact the Income Tax Division of the State Department of Revenue for details. Even if your state is not listed, call them up, because new legislation is appearing very quickly in this area.

Property and Sales Tax

Installing a solar system would normally increase the value of your property and hence the property tax payable. This additional cost on a solar investment has been removed or reduced in twenty-nine states. Many states have also excluded solar equipment from sales tax.

Grants and Loans

Low interest loans, or, in some cases grants, have been made available in the following states:

CALIFORNIA
Dept. of Housing and Community Development. (916) 445.4728.

ILLINOIS
Dept. of Business and Economic Development (217) 782.7500.

IOWA
IOWA Housing Finance Authority
(515) 281.4058.

MAINE
Office of Energy Resources
(207) 289.2196.

MASSACHUSETTS
Local Bank or Credit Union.

MONTANA
Public Service Commission
(406) 449.3008.

OREGON
Oregon Dept. of Energy
(503) 378.4128.

TENNESSEE
Tenn. Housing Development Authority
(6155) 741.3023.

Remember that, as mentioned above, claiming a government subsidized loan will decrease the amount of federal tax credit you can claim.

SWIMMING POOL HEATERS

For six months of the year, swimming pool water does not have to be heated any more than about 20°F above the air temperature. This means that swimming pool solar collectors can operate at low temperatures, losing very little of the heat they absorb to their surroundings, and thereby achieving high efficiencies. Swimming pool collectors raising the water temperature by only 2 or 3°F at each pass, usually operate at efficiencies in the range 70–80% compared with the typical 30–40% achieved by domestic water heaters. Not only do they yield more energy, but swimming pool collectors can cost much less than those designed to operate at higher temperatures. This is because the solar absorption plate, being only a few degrees above the air temperature, has such a low level of heat loss, it does not benefit significantly from layers of insulation or glazing. Indeed a transparent cover, which will reflect at least 16% of the solar radiation may actually reduce energy collection, except in cases where the absorber is exposed to frequent strong winds.

The absence of such a protective cover also enables the collector to operate on some occasions when there is no direct solar heat gain. This is possible because heat can be absorbed from the air which will sometimes be warmer than the water. System costs are further reduced by the fact that only the collection and circulation elements are additional, the pool itself fulfilling the function of heat storage. Finally swimming pools have the advantage that, unlike most energy demands, the demand follows closely the supply of solar energy. That is, the sunniest months are those in which swimming pools are most in use.

Value of Solar Heating Pools

The value of a solar heating system will of course vary not only with the size and location of the pool, but also with pattern of use to which it is put, and with the users' personal preferences. Some people for example especially value the bracing effect of a cold dip and you would have a hard time convincing them that any expenditure on heating is worthwhile.

To generalise one can say that there are three categories of pools:

i) Unheated pools
ii) Heated Pools
iii) Pools closing during summer

Unheated Pools

Only if you live some place like Florida can you expect a warm pool all year round without a fossil fuel fired heater. Unheated pools are generally used only during the summer. A solar system connected to such a pool would lengthen the swimming season by about two months. When compared with a fossil fuelled heating system, it would have the disadvantage that it would be unable to reheat the pool rapidly after dull periods and initial warming up in the spring would take several weeks. In addition the initial cost of a commercially installed solar system would perhaps be double that of a conventional boiler. The price of fuel however would whittle away the boiler's financial advantage in a short number of years. If the pool is to be used only during the summer months, there is no doubt that a solar installation would be the best choice.

Heated Pools

Heated pools use large amounts of expensive fuel the price of which will continue to rise in the years ahead. Legislation restricting such prolifigate use of energy already exists in California, Illinois and New York. In just about any form of analysis comparing the cost of a properly designed and installed swimming pool solar heating system with the savings it provides in the form of reduced fuel consumption, the system proves to be value for money. This is true both for the case where one has in hand the money to pay for the installation outright—the fuel savings outweigh the cash return which could have been realised through investing the money—and the case where the price of the installation has to be borrowed.

Pools Closing During Summer

This third category of pools is the only one where solar heaters may not be a worthwhile investment. There are schools where the swimming pools are unused during the summer vacation. This is of course a ridiculous waste of a valuable amenity which has probably been paid for in part or whole by public funds. Nevertheless it does happen and until education authorities and local clubs can cooperate to improve the situation, a careful study of the economic factors involved is required before committing large sums of money for the provision of solar heating. The problem is of course that the bulk of the savings that a solar system provides is contributed during the very months that schools are closed for holidays.

Insulating Covers

In all categories of swimming pools the first step in heating the water and reducing costs is to provide a pool cover. Once the pool has been initially heated at the beginning of the season, the main heating requirement is that of replacing heat lost through the pool surface. Heat is lost by radiation to the night sky and by convective currents and the evaporation of pool water. Compared with these factors, heat loss into the ground, in a sunken pool is insignificant. The cheapest form of cover is probably a layer of black PVC. An attractive alternative is a double skin of clear polythene with a matrix of air bubbles trapped between them. This has the advantage that it will not sink in the water and it allows solar radiation to pass through it. Pools which are more than a metre deep are reasonably good absorbers of solar radiation in their own right. A translucent plastic therefore will act as an insulative cover and still allow the pool to heat itself to a limited extent on sunny days. A family I once visited told me that they had tried various commercial covers in the past and then showed me the home-made version which they found preferable. They had stretched polythene (a U.V. resistant grade would be best) across three rectangular timber frames which spanned the breadth of the pool. This left an air space of about 1 ft (300 mm) above the water surface and hence it was possible to go swimming with the cover still on in rainy weather.

Collector Size

Most manufacturers recommend using an area of solar collectors equal to half the pool's surface area. It may be worth expanding this area if the collectors have to be mounted more than 25° away from south or if they are mounted at a very steep angle.

Collector Orientation

The ideal orientation is likely to be slightly west of south although the orientation of unglazed absorbers is even less critical than conventional panels which suffer from the increasing reflection at the glass surface when the radiation strikes at progressively more oblique angles. For the best results stay within 25° of south.

Collector Tilt

Generally a shallow tilt is desirable as the system will operate in summer when the sun is high in the sky. A steeper angle might be desirable in pools which may achieve quite satisfactory temperatures without much help during mid-summer; and also in combined systems (see end of this chapter) where the collectors are serving another energy demand in addition to the pool. All angles between horizontal and 60° are acceptable. They can even be mounted vertically although in such cases the area should be increased by over 50%.

Connecting the Solar System

There are many variations in the way a solar system can be connected to a swimming pool. The illustration indicates a simple method which involves breaking into the existing filtration circuit and thereby saving the expense of an additional pump. The existing pump may of course not be suitable to take the additional load of circulating through the collectors but this is unlikely as their flow resistance is small in comparison with that of the filter unit.

Following the circuit indicated, water is drawn off the pool from sump (1) and or the skimmer (2) by the pump (3). It is then passed through the filtration unit (4) and from there, it would normally be returned to the opposite end of the pool (5). When the solar system is added however, the water is first led to a three way, motorised valve (6). This is electrically controlled (8) and (9) so that when there is energy available in the solar panels, the stream is diverted to their bottom inlet (7). When there is no heat to be gained from the collectors, water is directed to the pool as before.

The collectors are connected in parallel and heated water is drawn off at the top corner (10) diagonally opposite the inlet. A drain-cock should be fitted in the remain-

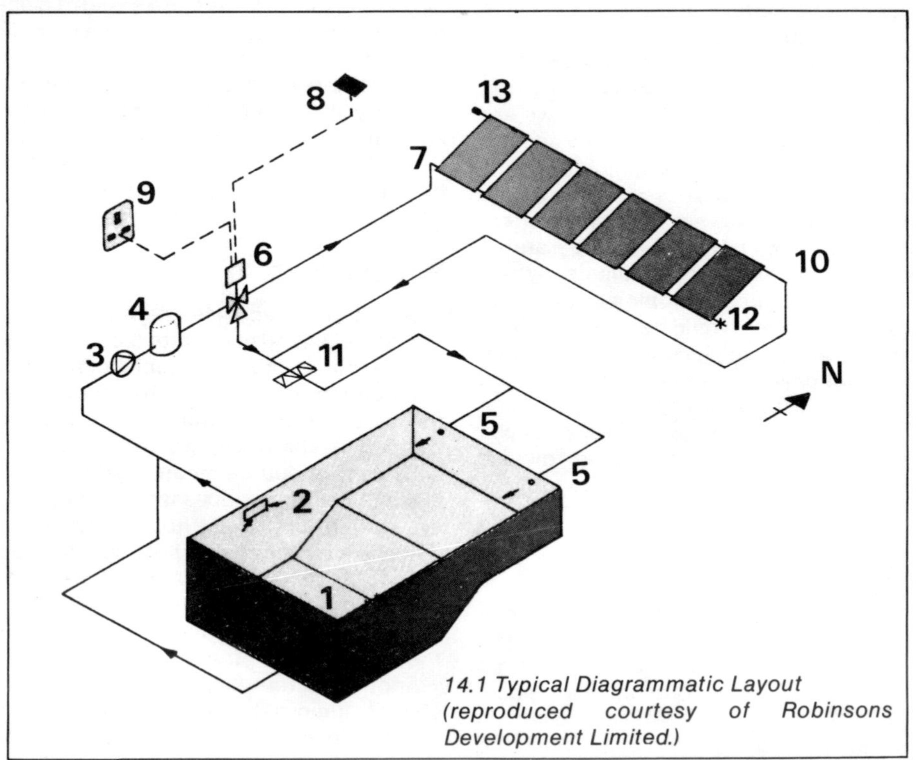

14.1 Typical Diagrammatic Layout (reproduced courtesy of Robinsons Development Limited.)

ing bottom corner (12) which should be the lowest point on the collector bank, and it is advisable to have an air vent at the highest point (13). Solar heated water is returned to the filtration circuit at a point before any additional heating unit(11).

It may be possible to economise on the capital cost of such a system by using two solenoid valves fitted in reverse in place of the expensive motorised valve. If a few hours operation each day would be sufficient for the job of filtration, the solar system operating costs could be reduced by installing a separate pump with a lower rating than the filtration pump.

Swimming Pool Collectors

The major differences between pool collectors and standard flat plate solar collectors are their lower operating temperature and the increased danger of corrosion.

As already stated, for low temperature applications such as pools, uninsulated and unglazed absorbers are usually adequate. This of course means that the exterior of the absorber plate is exposed to weathering. Furthermore, the swimming pool systems are almost invariably direct circuits. (At the low operating temperatures which are necessary to maintain maximum efficiency, a very large and expensive heat exchanger would be necessary in an indirect circuit.) This means that not only will the water passing through the collectors be aerated, but it will also contain the corrosive chemical compounds which are formed when small quantities of urine in the water interact with the pool's chlorination.

Given these added dangers extra care must be exercised in the choice of materials.

Copper and certain grades of stainless steel would probably be the most resistant to corrosion. They are however expensive. Aluminium and steel are liable to corrosion under such hostile conditions and we are therefore left with plastic collectors appearing as the best choice. The major problems encountered with plastics will be degradation due to ultra-violent radiation, softening and deformation at high temperatures and leaching of their stabilisers by hot water. Given these problems the best choice of plastics would be polypropylene and polycarbonate pigmented with carbon black, not only to increase absorption, but to resist the effect of U.V. radiation. A recent new design of swimming pool collector incorporates a plastic pipe coiled in a slab which can be laid as paving for the pool surrounds.

DIY Pool Collectors

Attempts have been made to make polypropylene sandwich type absorbers of the type shown in fig. 4.2 from a semi rigid wafer manufactured originally for packaging. The difficulty however comes in making a water tight bond between the open end of the wafer, and a header pipe. I have been told that this can be done with Dow Corning silicone sealant as long as pressure on the joint is low. I have not however seen such a joint tested.

A more proven DIY approach would be to use corrugated aluminium sheets, with an annodised or stoved on coating for corrosion prevention, and construct a trickle type collector as described in chapter 4 and chapter 15. Due to the high evaporation losses which would occur with an open system such as this, it would be advisable to enclose the absorber with a sealed transparent cover.

A closed version of the trickle collector which would not need a transparent cover has been described by Dr. Cleland McVeigh in his book 'Sun Power'. It uses a polythene/air bubble packaging material, mounted on a ply wood sheet in place of the corrugated metal. The air bubble matrix distributes the flow of water which is released by a header pipe along the upper edge. A sheet of black butyl is then spread across the assembly and sealed at the edges except along the bottom where the heated water is collected.

Finally it is worth mentioning that the solar absorber can be an object of beauty. It is possible to raise the pool temperature by cascading water over a series of black painted concrete terraces at the pool side.

Pump Control

Electronic differential temperature controllers as described in chapter 5 can be used but for swimming pool applications they must be more sensitive than usual due to the very low temperature differences at which it is still worth circulating. Greater sensitivity of course means higher cost.

Some control is necessary however as a system left running 24 hours a day would actually have a net effect of cooling a pool due to radiation to the night sky. The cheapest automatic control would be a time switch which turns the system on in the morning and off at night.

Another alternative which would be cheaper than a differential controller, is a device designed and manufactured by Robinsons Developments. This switches on the pump when the sol-air temperature reaches a pre-determined level. 'Sol-air' indicates a composite reading of radiation, humidity, wind, and air temperature. The device itself is quite simple though, consisting of a thermostat encased in a black metal box and placed in front of a polished metal reflector.

Combined Systems

A large bank of solar collectors serving a summer load in the form of a swimming pool, may be able to make a useful contribution to a space heating load during the other half of the year. Similarly, a domestic water heating installation will often collect more energy in summer than it needs and such surplus could be diverted to a swimming pool.

Where pools are sited close to buildings, such combined systems would make sense.

One would have to rethink the questions of glazing, insulation and tilt angles, bearing in mind which system is considered to have most priority. Sophisticated electronic controls would play an important role in the success of such a system.

Heat Pumps

Heat pumps are devices which extract heat from a source which is at a temperature, and upgrade this heat to deliver it at higher temperatures. Refrigerators do this. They extract heat from their interiors, which start out at room temperature, and expel the heat in the form of warm air at the rear.

To carry out this operation, the heat pump requires an energy input, usually electrical. To be worthwhile of course, the device must 'pump out' more energy than it consumes. Typically they can deliver about 3kWh for every 1kWh of electricity they use. The proportion will become less favourable however as the temperature difference between the heat source, and the delivered heat increases.

Pools & Heat Pumps

This is where swimming pools may prove useful. Normally the heat source is the open air, or a stream, or the ground itself. A swimming pool, particularly one connected to a solar system, is likely to be at a higher temperature than the air or a stream and will be more conveniently tapped than the ground. The heat pump will also complement the performance of the solar collectors. By reducing the pool's winter temperature, it will lower the collector's operating temperature thereby increasing its efficiency and extending its period of useful functioning.

If a new pool is being constructed, it is worth siting it near a building and installing ducts which could be used for connecting a heat pump. Even if such a system is not immediately desirable, such preparations would open up options for the future.

Solar Ponds

It has been mentioned that the pool itself can collect solar energy. Much of the absorbed heat is however raised to the surface by convection currents and there lost to the air. The efficiency of an energy collecting pool can be greatly increased if the convection currents can be suppressed. The means of achieving this is to dissolve salt in the water in such a way as to form a gradient of salinity, with a saturated solution at the pool bottom and much less salty at the surface. In such a pond, absorbed heat would remain at the bottom and high temperatures could build up. The most famous example of such a solar pond is the one opened in 1979 on the shores of the Dead Sea in Israel which has been connected to a low temperature turbine and produces 150 kW of electricity.

Such schemes are often described as being suitable only for lower latitudes. However a theoretical study by Bryant and Colbeck published in the scientific journal 'Solar Energy', suggests that even in cloudy London at latitude 52° N. a solar pond with a surface area equal to the floor area of a house, could heat that house throughout winter at a cost comparable with that of gas if the installation was paid for on a twenty year mortgage. The pond in the study was three metres (10ft) deep, an upper layer of salt saturated water two metres deep, and underneath a separating membrane one metre of plain water.

Will swimming pools in the future no longer be a recreational luxury but a necessary part of keeping warm in a fuel-deprived world? Shall we see new houses being built in groups around artificial ponds which not only keep the children happy during summer, but also soak up the summer sunshine and waste heat from the buildings to return it during cold spells? These questions cannot be answered now but they may be answered sooner than you think. Which would you opt for if given the choice between a backyard solar pond and a central heating boiler?

Appendix (ii) Books

There are scores of books available these days on solar energy. This is a very encouraging sign (for everyone except the authors of solar energy books). But it does make choosing difficult. These then are some of the books I have found most useful: they are grouped in two categories. First come the books directly relevent to solar water heating; then books which are related, but focused on more general topics.

Installation Guidelines for Solar Domestic Hot Water (1977)
Franklin Research Center, published by the US Dept. of Housing and Urban Development (HUD-PDR-407) and available from Superintendent of Documents, US Govt. Printing Office, Washington DC 20402. ($4.00)
An excellent guide for professionals and enthusiasts who are involved in installation of solar systems. Lots of practical advice based on lessons learned from the federal demonstration program on solar water heating.

HUD Intermediate Minimum Property Standards for Solar Heating 2nd Domestic Hot Water Systems — Vol. 5 (IMPS) 1977
US Dept. of Housing and Urban Development available from US Govt Printing Office (Stock No. 4930.2) $12.00
Lays down the standards that were required for participants in federally funded programs. A large section is devoted to presenting the f-chart method of calculating solar contribution to energy requirements.

Uniform Solar Energy Code 1976
International Association of Plumbing and Mechanical Officials, 5032 Alhambra Ave., Los Angeles, CA 90032 ($8.50)
Suggested specification for safe and functional solar systems. Includes lengthy section on collector sizing calculations.

New Inventions in Low-Cost Solar Heating
William Shurcliff, Brick House, Andover, Mass 01810 ($12.00). Aimed at introducing better and cheaper solar systems, this is a large and stimulating collection of solar bright ideas.

How to Build a Passive System Solar Water Heater 1976
Horace McCracken, Rt.1, Box 417, Alpine CA92001 and,
Breadbox Water Heaters
Zomeworks Corp., PO Box 712 Albuquerque NM87103

These are both sets of plans to build the breadbox type of solar collectors described in chapter 4 of this book.

Solar Systems in B.C. 1979
R.Bryenton, K.Cooper, C.Mattock and T.Lyster, published by BC Hydro, Vancouver.
A Solar Water Heater Workshop Manual
Ecotope Group, 2332 East Madison, Seattle, WA 98112.

These books both describe the construction and installation of a low cost collector consisting of copper tubes clamped with wire onto a corrugated metal sheet. The books benefit from a breadth of experience gained in many workshop sessions.

Books of Related Interest

Direct Use of the Sun's Energy
Farrington Daniels
Ballantine Paperback, 1964
Perhaps the most popular book written on solar energy, it presents an over-view of all lines of research aimed at harnessing the sun's energy. It is richly annotated with references for those who wish to follow up particular interests in more detail, but written in simple language which can be easily assimiliated by non-specialists.

Sun Power
J.C.McVeigh
Pergamon, 1977
A more up to date over-view than Farrington Daniels, this is an excellent starting point for the student who wishes to delve into the many applications of solar energy. It draws successfully on the author's personal experience with low temperature applications and his familiarity with international research.

Sun Spots
Steve Baer
Baer is well known for his innovative solar devices and his book is equally innovative. Sections on heat exchangers and rock pile storage are interlaid with conspiracy theories and visions of state intervention to interfere with individuals' access to sunlight. Sometimes fictional sequences serve as vivid illustrations of the factual information. The story of the petrol-drinking addict for example, is a strong reminder of the importance of energy in our daily lives, and its role as a unifying principle.

Solar Energy for Man
B.J.Brinkworth
The Compton Press, 1972.
If you become really intrigued by solar energy, it is well worth reading this book. It is more a student's text than the other books mentioned, but care has been taken to jog the memories of those who must struggle to remember their school physics. The effort is amply rewarded by the satisfaction of establishing a clear theoretical understanding of the various physical processes which take place between nuclear explosions in the heart of the sun and your bathful of hot water.

The Solar Home Book
Bruce Anderson
This is probably the most readable account of how to use solar energy for heating the home. It places a welcome emphasis on simple passive techniques whilst still detailing the more expensive mechanical solutions using collector/storage circulation loops. Written primarily for those who design or build houses, the layman is catered for by features on DIY projects and by graphical explanations of the technical terms and concepts which are mentioned.

Designing and Building a Solar House
Donald Watson, Garden Way, Charlotte VT.
One of my favourite solar books. Lots of ideas about how to fit a solar system into existing houses.

The Passive Solar Energy Book 1979
Ed. Mazria, Rodale Press, Emmaus Penn.
Beautifully illustrated guide to the most cost effective ways to utilise solar heat. Full of rules of thumb to assist in the early stages of design.

Solar Greenhouses (1976 updated 1979)
Bill Yanda and Rick Fisher.
John Muir Publications, Santa Fe, New Mexico.
If you have a south facing wall on your house and a piece of ground in front of it, a solar greenhouse is just about the most sensible thing you could do for yourself. And this book shows how to build, plant it, and estimate your benefits.

The Readers Digest Repair Manual — A Complete Guide to Home Maintenance
The Readers Digest Co.
With its clear and abundant illustrations this book can help you understand how your house works and thus help you decide how to fit a solar system on to it.

The Practical Handbook of Plumbing and Heating 1969
Richard Day, Arco Publishing Co. New York.
Plumbing is the critical skill for solar water heating installation. If chapters 8 and 9 in the book you are holding do not give you enough confidence, try this book by Richard Day.

Soft Energy Paths
Amory Lovins
Penguin Books, 1977.
A welcome reassurance to society dithering on the brink of change. By means of a rational appraisal of all our potential energy resources, from the "hard" solutions like nuclear power, to the "soft technologies" utilising the sun, wind and biomass conversion, Lovins demonstrates that the soft paths are not only safer, but also cheaper. Those who have kept themselves outside the arguments on energy policy because of the tendency for debates to become lost in technical intricacies, will be pleased to read Lovins' view that: "The basic issues in energy strategy, far from being too complex and technical for ordinary people to understand, are on the contrary too simple and political for the experts to understand."

Appendix (iii) Organisations

National Solar Heating and Cooling Information Center
P.O. Box 1607
Rockville
MD 20850
(800) 523-2929
Pennsylvania: (800) 462-4983
Alaska & Hawaii: (800) 523-4700 or your local IRS Office.
This is the key address for solar information. You can call them on their toll-free number to find out any other address.

Regional Centers:

Northeast Solar Energy Center (NESEC)
470 Atlantic Avenue
Boston, Massachusetts 02110
Tel: (617) 292-9250

Southern Solar Energy Center (SSEC)
61 Perimeter Park
Atlanta, Georgia 30341
Tel: (404) 458-8765

Western Sun
Pioneer Park Bldg., Suite 800
715 S.W. Morrison
Portland, Oregon 97205
Tel: (503) 221-2437

Mid-American Solar Energy Complex (MASEC)
8140 26th Avenue South
Minneapolis, Minnesota
Tel: (612) 853-0400

These regional centers have been set up to support the development of solar energy by the US Dept. of Energy.

American Section of the International Solar Energy Society
Research Institute for Advanced Technology
US Highway 190 West
Killeen, Texas 76541

It is worth joining the AS-ISES if only for the free subscription to Solar Age Magazine—the enjoyable way to keep up to date.

Solar Usage Now (S.U.N.)
Box 306
Bascom, OH 44809

For equipment and materials this is a good place to try. SUN sell a thick catalog for $5.00 and sell goods mail order. Everthing from solar tanks to badges.

Telephone numbers marked * below are toll-free in the state only.

Alabama Solar Energy Center
University of Alabama/Huntsville
Huntsville, AL 35807
*(800) 572-7226
(205) 895-6361

Alabama Department of Energy
3734 Atlantic Highway
Montgomery, AL 36130
(205) 832-5010

Alaska Department of Commerce & Economic Development
Division of Energy & Power Development
7th Floor
338 Denali Street
Anchorage, AK 99501
(907) 276-0508

Arizona Solar Energy Research Commission
700 West Washington, Room 502
Phoenix, AZ 85007
*(800) 352-5499
(602) 255-3303

Arkansas Department of Energy
300 Kavanaugh
Little Rock, AR 72205
*(800) 482-1122
(501) 371-1370

California Energy Commission
1111 Howe Avenue, Mail Station 70
Sacramento, CA 95825
*(800) 852-7516
(916) 920-6019

Colorado Office of Energy Conservation
1600 Downing Street
Denver, CO 80218

Connecticut Office of Policy & Management
Energy Division
80 Washington Street
Hartford, CT 06115
Tel. *(800) 842-1648
(203) 566-2800

Delaware Governor's Energy Office
56, The Green
Dover, DE 19901
Tel. *(800) 282-8616
(302) 678-5644

Florida Solar Energy Center
300 State Road, 401
Cape Canaveral, FL 32920
Tel. (305) 783-0300

Governor's Energy Office
301 Bryant Building
Tallahassee, FL 32301
(904) 488-6764

Georgia Office of Energy Resources
Room 615
270 Washington Street, S.W.
Atlanta, GA 30334
(404) 656-5176

Hawaii State Energy Office
Department of Planning & Economic Development
P.O. Box 2359
Honolulu, HI 96804
Tel. (808) 548-4150
(808) 548-4080 (on Oahu)

Idaho State Office of Energy
State House
Boise, ID 83720
Tel. (208) 334-3800

Institute of Natural Resources
325 West Adams, Room 300
Springfield, IL 62706
Tel. (217) 785-2800

Indiana Energy Group
440 North Meridan
Indianapolis, IN 46204
Tel. (317) 232-8940

Iowa Solar Office
Energy Policy Council
Capitol Complex
Des Moines, IA 50319
Tel. (800) 532-1114
(515) 281-8071

Kansas Energy Office
241 West 6th Street
Topeka, KS 66603
*(800) 432-3537
(913) 296-2496

Kentucky Department of Energy
P.O. Box 11888
Lexington, KY 40578
(606) 252-5535

Louisiana Department of Natural Resources
Research & Development Division
P.O. Box 44156
Baton Rouge, LA 70804
(504) 342-4594

Maine Office of Energy Resources
State House, Station 53
Augusta, ME 04333
(207) 289-3811

Maryland Energy Policy Office
Suite 903
301 West Preston Street
Baltimore, MD 21201
*(800) 494-5903
(301) 383-6810

Massachusetts Office of Energy Resources
73 Tremont Street
Boston, MA 02108
*(800) 922-8265
(617) 727-7297

Michigan Energy Administration
6520 Mercantile Way, Suite 1
Lansing, MI 48910
*(800) 292-4704
(517) 373-6430

Minnesota State Energy Agency
980 American Center Building
150 East Kellog
St. Paul, MN 55101
*(800) 652-9747
(612) 296-5120

Mississippi Energy & Transportation Department
Watkins Building
570 George Street
Jackson, MS 39202
*(800) 932-2310
(601) 961-4403

Missouri Department of Energy
P.O. Box 176
1915 South Ridge Drive
Jefferson City, MO 65102
*(800) 392-0717
(314) 751-4000

Montana Department of Natural Resources and Conservation
32 South Ewing
Helena, MT 59601
(406) 449-4624

Nebraska Solar Office
W-191 Nebraska Hall
University of Nebraska
Lincoln, NE 68588
(402) 472-3414

Nebraska Energy Office
State Capitol, 9th Floor
Lincoln, NE 68509
(402) 471-2867

Nevada Department of Energy
400 West King Street
Carson City, NV 89710
*(800) 992-0900
(702) 885-5157

New Hampshire Governor's Council on Energy
2 1/2 Beam Street
Concord, NH 03301
*(800) 852-3466
(603) 271-2711

New Jersey Department of Energy Office of Alternative Technology
101 Commerce Street
Newark, NJ 07102
*(800) 492-4242
(201) 648-6293

New Mexico Energy & Minerals Department
Energy Conservation & Management Division
P.O. Box 2270
Sante Fe, NM 87501
*(800) 432-6782
(505) 827-5621
(505) 827-2386

New York State Energy Office
Agency Building #2
Empire State Plaza
Albany, NY 12223
*(800) 342-3722
(518) 474-7016

North Carolina Department of Commerce
Energy Division
430 North Salisbury Street
Raleigh, NC 27611
*(800) 662-7131
(919) 733-4490

North Dakota State Solar Office
1533 North 12th Street
Bismarck, ND 58501
(701) 224-2250

Ohio Solar Office
30 East Broad Street
34th Floor
Columbus, OH 43215
*(800) 282-9234
(614) 466-8277

Oklahoma Department of Energy
4400 North Lincoln Boulevard
Suite 251
Oklahoma City, OK 73105
(405) 521-3941

Oregon Department of Energy
Room 102
Labor & Industries Building
Salem, OR 97310
*(800) 452-7813
(503) 378-4040

Pennsylvania Governor's Energy Council
1625 North Front Street
Harrisburg, PA 17102
*(800) 882-8400
(717) 783-9981

Puerto Rico Office of Energy
Energy Information Program
P.O. Box 41089, Minillas Station
Santource, PR 00940
(809) 727-8877

Rhode Island Governor's Energy Office
80 Dean Street
Providence, RI 02903
Providence, RI 02903
(401) 277-3773
(instate collect calls accepted)

South Carolina Division of Energy Resources
SCN Center Suite 1130
1122 Lady Street
Columbia, SC 29201
*(800) 922-5310
(803) 758-2050

South Dakota Office of Energy Policy
Capital Lake Plaza
Pierre, SD 57501
*(800) 592-1865
(605) 773-3603

Tennessee Energy Authority
Suite 707, Capitol Boulevard Building
Nashville, TN 37219
*(800) 642-1340
(615) 741-6671

Texas Energy and Natural Resources
1800 San Jacinto
5th Floor
Austin, TX 78701
(512) 475-5407

Utah Energy Office
231 East 400 South
Empire Building, Suite 101
Salt Lake City, UT 84111
*(800) 662-3633
(801) 533-5424

Vermont State Energy Office
State Office Building
Montpelier, VT 05602
*(800) 642-3281
(802) 828-2393

Virginia Division of Energy
310 Turner Road
Richmond, VA 23225
*(800) 552-3831
(804) 745-3245

Washington State Energy Office
400 East Union Street
1st Floor
Olympia, WA 98504
(206) 754-0700

West Virginia Fuel and Energy Office
126 1/2 Greenbrier Street
Charleston, WV 25311
*(800) 642-9012
(304) 348-8860

Wisconsin Division of State Energy
101 South Webster
8th Floor
Madison, WI, 53702
*(800) 362-9696
(608) 266-9861

Wyoming Energy Conservation Office
320 West 25th Street
Capitol Hill Building
Cheyenne, WY 82002
(307) 777-7131

Virgin Islands Energy Office
P.O. Box 2996
St. Thomas
U.S. Virgin Islands 00801
(809) 774-6726

INDEX